ARE **4.0**

2010

Building Design & Construction Systems

Study Guide   Caleb Hornbostel & Lester Wertheimer

KAPLAN

AE EDUCATION

President: Dr. Andrew Temte

Chief Learning Officer: Dr. Tim Smaby

Vice President of Engineering Education: Dr. Jeffrey Manzi, PE

Senior Product Manager: Brian O'Connor

BUILDING DESIGN & CONSTRUCTION SYSTEMS STUDY GUIDE: 2010 EDITION

© 1986, 1994, 1997 by Architectural License Seminars, Inc.

© 2009 by Dearborn Financial Publishing, Inc.®

Published by Kaplan Architecture Education

1905 Palace Street

La Crosse, WI 54603

800-420-1429

www.kaplanarchitecture.com

Printed in the United States of America.

09   10   11   10 9 8 7 6 5 4 3 2 1

ISBN: 1-4277-8680-1
PPN: 1629-3102

CONTENTS

WELCOME

Thank you for choosing Kaplan AE Education for your ARE study needs. We offer updates annually to keep abreast of code and exam changes and to address any errors discovered since the previous update was published. We wish you the best of luck in your pursuit of licensure.

ARE OVERVIEW

Since the State of Illinois first pioneered the practice of licensing architects in 1897, architectural licensing has been increasingly adopted as a means to protect the public health, safety, and welfare. Today, the United States and Canadian provinces require licensing for individuals practicing architecture. Licensing requirements vary by jurisdiction; however, the minimum requirements are uniform and in all cases include passing the Architect Registration Examination (ARE). This makes the ARE a required rite of passage for all those entering the profession, and you should be congratulated on undertaking this challenging endeavor.

Developed by the National Council of Architectural Registration Boards (NCARB), the ARE is the only exam by which architecture candidates can become registered in the United States or Canada. The ARE assesses candidates' knowledge, skills, and abilities in seven different areas of professional practice, including a candidate's competency in decision making and knowledge of various areas of the profession. The exam also tests competence in fulfilling an architect's responsibilities and in coordinating the activities of others while working with a team of design and construction specialists. In all jurisdictions, candidates must pass the seven divisions of the exam to become registered.

The ARE is designed and prepared by architects, making it a practice-based exam. It is generally not a test of academic knowledge, but rather a means to test decision-making ability as it relates to the responsibilities of the architectural profession. For example, the exam does not expect candidates to memorize specific details of the building code, but requires them to understand a model code's general requirements, scope, and purpose, and to know the architect's responsibilities related to that code. As such, there is no substitute for a well-rounded internship to help prepare for the ARE.

4.0 Exam Format

The seven ARE 4.0 divisions are outlined in the table below.

DIVISION	QUESTIONS	VIGNETTES
Building Design & Construction Systems	85	Accessibility/ Ramp Roof Plan Stair Design
Building Systems	95	Mechanical & Electrical Plan
Construction Documents & Services	100	Building Section
Programming, Planning & Practice	85	Site Zoning
Schematic Design	-	Building Layout Interior Layout
Site Planning & Design	65	Site Design Site Grading
Structural Systems	125	Structural Layout

The exam presents multiple-choice questions individually. Candidates may answer questions, skip questions, or mark questions for further review. Candidates may also move backward or forward within the exam using simple on-

ARCHITECTURAL HISTORY

Questions pertaining to the history of architecture appear throughout the ARE. The prominence of historical questions will vary not only by division but also within different versions of the exam for each division. In general, however, history tends to be lightly tested, with approximately three to seven history questions per division, depending upon the total number of questions within the division. One aspect common to all the divisions is that whatever history questions are presented will be related to that division's subject matter. For example, a question regarding Chicago's John Hancock Center and the purpose of its unique exterior cross bracing may appear on the Structural Systems exam.

Though it is difficult to predict how essential your knowledge of architectural history will be to passing any of the multiple-choice divisions, it is recommended that you refer to a primer in this field—such as Kaplan's *Architectural History*—before taking each exam, and that you keep an eye out for topics relevant to the division for which you are studying. It is always better to be overprepared than taken by surprise at the testing center.

screen icons. The vignettes require candidates to create a graphic solution according to program and code requirements.

Actual appointment times for taking the exam are slightly longer than the actual exam time, allowing candidates to check in and out of the testing center. All ARE candidates are encouraged to review NCARB's *ARE Guidelines* for further detail about the exam format. These guidelines are available via free download at NCARB's Web site (*www.ncarb.org*).

Exam Format

It is important for exam candidates to familiarize themselves not only with exam content, but also with question format. Familiarity with the basic question types found in the ARE will reduce confusion, save time, and help you pass the exam. The ARE contains three basic question types. The first and most common type is a straightforward multiple-choice question followed by four choices (A, B, C, and D). Candidates are expected to select the correct answer.

This type of question is shown in the following example.

> Which of the following cities is the capital of the United States?
>
> **A.** New York
>
> **B. Washington, D.C.**
>
> **C.** Chicago
>
> **D.** Los Angeles

The second type of question is a negatively worded question. In questions such as this, the negative wording is usually highlighted using all caps, as shown below.

> Which of the following cities is *NOT* located on the west coast of the United States?
>
> **A.** Los Angeles
>
> **B.** San Diego
>
> **C.** San Francisco
>
> **D. New York**

The third type of question is a combination question. In a combination question, more than one choice may be correct; candidates must select from combinations of potentially correct choices. An example of a combination question is shown below.

Which of the following cities are located within the United States?

I. New York

II. Toronto

III. Montreal

IV. Los Angeles

A. I only

B. I and II

C. II and III

D. I and IV

The single most important thing candidates can do to prepare themselves for the vignettes is to learn to proficiently navigate NCARB's graphic software. Practice software can be downloaded free of charge from their Web site. Candidates should download it and become thoroughly familiar with its use.

Recommendations on Exam Division Order

NCARB allows candidates to choose the order in which they take the exams, and the choice is an important one. While only you know what works best for you, the following are some general considerations that many have found to be beneficial:

1. The Building Design & Construction Systems and Programming, Planning & Practice divisions are perhaps the broadest of all the divisions. Although this can make them among the most intimidating, taking these divisions early in the process will give a candidate a broad base of knowledge and may prove helpful in preparing for subsequent divisions. An alternative to this approach is to take these two divisions last, because you will already be familiar with much of their content. This latter approach likely is most beneficial when you take the exam divisions in fairly rapid succession so that details learned while studying for earlier divisions will still be fresh in your mind.

2. The Construction Documents & Services exam covers a broad range of subjects, dealing primarily with the architect's role and responsibilities within the building design and construction team. Because these subjects serve as one of the core foundations of the ARE, it may be advisable to take this division early in the process, as knowledge gained preparing for this exam can help in subsequent divisions.

3. Take exams that particularly concern you early in the process. NCARB rules prohibit retaking an exam for six months. Therefore, failing an exam early in the process will allow the candidate to use the waiting period to prepare for and take other exams.

EXAM PREPARATION

Overview

There is little argument that preparation is key to passing the ARE. With this in mind, Kaplan has developed a complete learning system for each exam division, including study guides, question-and-answer handbooks, mock exams, and flash cards. The study guides offer a condensed course of study and will best prepare you for the exam when utilized along with the other tools in the learning system. The system is designed to provide you with the general background necessary to pass the exam and to provide an indication of specific content areas that demand additional attention.

In addition to the Kaplan learning system, materials from industry-standard documents may prove useful for the various divisions. Several of these sources are noted in the "Supplementary Study Materials" section on the following page.

Understanding the Field

The study of construction materials and methods is useful to architects in several ways. First, one must be aware of construction industry standards to determine what is safe and practical. Second, one must be able to communicate design ideas through detailed drawings to those in the field who will actually perform the work. Finally, one must recognize the limitations of skilled craftsmen while appreciating their ability to translate mere lines on paper into a real building.

In the past several years, building technology has become enormously complex and specialized, and new developments tend toward making past methods obsolete. While keeping up with all recent developments would be unrealistic, the goal should be to gain an understanding of the basic materials used in construction and to learn how these are put together to achieve strong, useful, and beautiful structures.

Understanding the Exam

The Building Design & Construction Systems exam is among the broadest of all the ARE divisions. Because many of the subjects covered span other divisions, it is widely considered one of the more difficult and intimidating for which to prepare. However, the exam tends to focus on the general principles and properties of common building materials while avoiding specific data and calculations.

Many have found that this division places a particular emphasis on sustainable design, the properties of materials, and definitions of various materials and methods. The Kaplan flash cards should prove especially helpful in mastering the definitions.

Preparation Basics

The first step in preparation should be a review of the exam specifications and reference materials published by NCARB. These statements are available for each of the seven ARE divisions to serve as a guide for preparing for the exam. Download these statements and familiarize yourself with their content. This will help you focus your attention on the subjects on which the exam focuses.

Prior CAD knowledge is not necessary to successfully complete vignettes. In fact, it's important for candidates familiar with CAD to realize they will experience significant differences between CAD and the drawing tools used on the exam.

Though no two people will have exactly the same ARE experience, the following are recommended best practices to adopt in your studies and should serve as a guide.

Set aside scheduled study time.
Establish a routine and adopt study strategies that reflect your strengths and mirror your approach in other successful academic pursuits. Most importantly, set aside a definite amount of study time each week—just as if you were taking a lecture course—and carefully read all of the material.

Take—and retake—quizzes.
After studying each lesson in the study guide, take the quiz found at its conclusion. The quiz questions are intended to be straightforward and objective. Answers and explanations can be

found at the back of the book. If you answer a question incorrectly, see if you can determine why the correct answer is correct before reading the explanation. Retake the quiz until you answer every question correctly and understand why the correct answers are correct.

Identify areas for improvement.
The quizzes allow you the opportunity to pinpoint areas where you need improvement. Reread and take note of the sections that cover these areas and seek additional information from other sources. Use the question-and-answer handbook and online test bank as a final tune-up for the exam.

Take the final exam.
A final exam designed to simulate the ARE follows the last lesson of each study guide. Answers and explanations can be found on the pages following the exam. As with the lesson quizzes, retake the final exam until you answer every question correctly and understand why the correct answers are correct.

Use the flash cards.
If you've purchased the flash cards, go through them once and set aside any terms you know at first glance. Take the rest to work, reviewing them on the train, over lunch, or before bed. Remove cards as you become familiar with their terms until you know all the terms. Review all the cards a final time before taking the exam.

Practice using the NCARB software.
Work through the practice vignettes contained within the NCARB software. You should work through each vignette repeatedly until you can solve it easily. As your skills develop, track how long it takes to work through a solution for each vignette.

Supplementary Study Materials

In addition to the Kaplan learning system, materials from industry-standard sources may prove useful in your studies. Candidates should consult the list of exam references in the NCARB guidelines for the council's recommendations and pay particular attention to the following publications, which are essential to successfully completing this exam:

- International Code Council (ICC) *International Building Code*
- *Standard on Accessible and Usable Buildings and Facilities* (ICC/ANSI A117.1-98)
- National Fire Protection Association *Life Safety Code* (NFPA 101)
- American Institute of Architects B141-1997 *Standard Form of Agreement Between Owner and Architect*
- American Institute of Architects A201-1997 *General Conditions of the Contract for Construction*
- American Institute of Steel Construction *Manual of Steel Construction: Allowable Stress Design*, Ninth Edition

Test-Taking Advice

Preparation for the exam should include a review of successful test-taking procedures—especially for those who have been out of the classroom for some time. Following is advice to aid in your success.

Pace yourself.
Each division allows candidates at least one minute per question. You should be able to comfortably read and reread each question and fully understand what is being asked before answering. Each vignette allows candidates ample time to complete a solution within the time allotted.

Read carefully.

Begin each question by reading it carefully and fully reviewing the choices, eliminating those that are obviously incorrect. Interpret language literally, and keep an eye out for negatively worded questions. With vignettes, carefully review instructions and requirements. Quickly make a list of program and code requirements to check your work against as you proceed through the vignette.

Guess.

All unanswered questions are considered incorrect, so answer every question. If you are unsure of the correct answer, select your best guess and/or mark the question for later review. If you continue to be unsure of the answer after returning to the question a second time, it is usually best to stick with your first guess.

Review difficult questions.

The exam allows candidates to review and change answers within the time limit. Utilize this feature to mark troubling questions for review upon completing the rest of the exam.

Reference material.

Some divisions include reference materials accessible through an on-screen icon. These materials include formulas and other reference content that may prove helpful when answering questions in these divisions. Note that candidates may *not* bring reference material with them to the testing center.

Best answer questions.

Many candidates fall victim to questions seeking the "best" answer. In these cases, it may appear at first glance as though several choices are correct. Remember the importance of reviewing the question carefully and interpreting the language literally. Consider the following example.

Which of these cities is located on the east coast of the United States?

A. Boston

B. Philadelphia

C. Washington, D.C.

D. Atlanta

At first glance, it may appear that all of the cities could be correct answers. However, if you interpret the question literally, you'll identify the critical phrase as "on the east coast." Although each of the cities listed is arguably an "eastern" city, only Boston sits on the Atlantic coast. All the other choices are located in the eastern part of the country but are not coastal cities.

Style doesn't count.

Vignettes are graded on their conformance with program requirements and instructions. Don't waste time creating aesthetically pleasing solutions and adding unnecessary design elements.

ACKNOWLEDGMENTS

This course was written by Caleb Hornbostel, a noted architect, teacher, and author. Mr. Hornbostel has taught at Pratt Institute, Columbia, Cornell, NYU, and Temple University.

The material in Lesson 9 was written by John A. Raeber, FAIA, FCSI, CCS. Mr. Raeber has been teaching classes for this division for 30 years.

The material in Lesson 10 was written by Jonathan Boyer, AIA. Mr. Boyer is a principal of the firm Farr Associates in Chicago, Illinois. He is a graduate of the University of Pennsylvania (BA) and Yale University (MArch). His practice has focused on sustainable design and environmental planning for more than 30 years with projects throughout the United States.

Passages addressing the history of building construction materials and their applications were developed for the 2009 edition by Dr. Shelley Roff. Dr. Roff is an associate professor in the School of Architecture at the University of Texas at San Antonio. As an architectural historian, her expertise is in development of civic architecture and the construction trades in medieval Spain and colonial America. She is the director of URBIS, a virtual simulation research lab for the study of contemporary and historical urban environments. Before receiving her PhD from Brown University, she worked as an architect in San Francisco and Boston. Her architectural education was formed with a BED from Texas A&M University and Master of Architecture degree from the University of California at Berkeley.

This introduction was written by John F. Hardt, AIA. Mr. Hardt is vice president and senior project architect with Karlsberger, an architecture, planning, and design firm based in Columbus, Ohio. He is a graduate of Ohio State University (MArch).

ABOUT KAPLAN

Thank you for choosing Kaplan AE Education as your source for ARE preparation materials. Whether helping future professors prepare for the GRE or providing tomorrow's doctors the tools they need to pass the MCAT, Kaplan possesses more than 50 years of experience as a global leader in exam prep and educational publishing. It is that experience and history that Kaplan brings to the world of architectural education, pairing unparalleled resources with acknowledged experts in ARE content areas to bring you the very best in licensure study materials.

Only Kaplan AE offers a complete catalog of individual products and integrated learning systems to help you pass all seven divisions of the ARE. Kaplan's ARE materials include study guides, mock exams, question-and-answer handbooks, video workshops, and flash cards. Products may be purchased individually or in division-specific learning systems to suit your needs. These systems are designed to help you better focus on essential information for each division, provide flexibility in how you study, and save you money.

To order, please visit *www.kaplanarchitecture.com* or call 800-420-1429.

Part I

The Multiple-
Choice Exam

CONCRETE

INTRODUCTION

Concrete is a manufactured construction material that is a mixture of fine aggregate (sand); coarse aggregate (gravel or crushed rock); portland cement; and water.

The aggregates in concrete are the inert ingredients, while cement and water are the active ingredients. The aggregates, usually sand and gravel, are first mixed thoroughly with the cement. As water is added, a chemical reaction takes place between the water and cement that creates heat and causes the concrete to harden.

It is this chemical reaction, called *hydration*, rather than the drying out of the mix, that causes hardening of the concrete. In fact, during the curing process, concrete must be kept

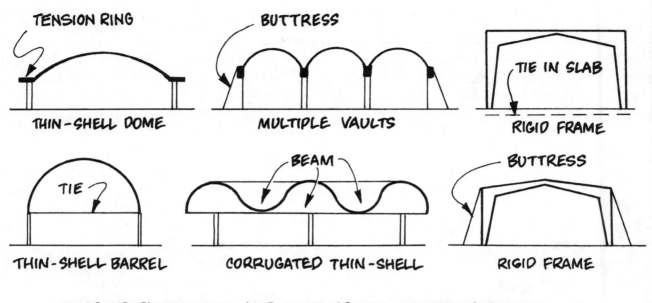

TYPES OF THIN-SHELL AND RIGID FRAME CONCRETE FORMS

Figure 1.1

moist for satisfactory hydration of the cement. Confirmation of this chemical process is the fact that concrete will harden just as well under water as in the air.

The use of concrete as a building material has greatly increased in recent years to a point where all types of structures, from simple bus shelters to multistory buildings, are constructed from essentially this one material. Concrete is used for foundations, floors, columns, walls, beams, and roofs, for precast panels in floors and walls, in thin-shell structures, in rigid frames, in decorative cast shapes, and for bricks and blocks that are used in concrete masonry construction. Concrete, it would appear, is a material for all purposes and all times.

HISTORY

Discovery and Early Uses

The Greeks began experimenting with formulas for concrete as early as 1700 BCE. However,

nothing produced a material with the qualities needed for building construction until the Romans began mixing a lime mortar with volcanic ash from the slopes of Mount Vesuvius in the third century BCE. This pozzolan ash, a silica-alumina based mineral taking its name from the Pozzouli region in the Bay of Naples, has a unique characteristic: when mixed with lime and a small amount of water, a chemical reaction causes the ash to bond with the lime and harden at ordinary temperatures. The difference between pozzolan mortar and traditional lime mortar made with river sand was that the volcano had previously *calcined*, or heated, the mineral ash. Along parallel lines, modern portland cement is heated in a kiln and the resultant clinker is crushed into a powder.

The Roman engineer Vitruvius understood the key importance of this secret in making Roman concrete and dedicated an entire chapter to pozzolan in his architectural treatise, *De Architectura*. He also discusses other critical factors, such as using a minimum of water to create a stiff mortar, a "no slump" mixture, and special placement techniques. Roman laborers tamped this cement into the voids of rubble layers using special tools, a practice similar to present-day Roller Compacted Concrete (RCC) used in dam construction. Although Roman concrete used no reinforcing materials, its tremendous longevity—2,000 years and counting—is attributed to its low water content and close compaction.

Ancient writing describes some failures in concrete construction; however, by the first century CE this process had reached its zenith with a multitude of new structural forms and building types. The Romans discovered that this cement produced an even stronger concrete when allowed to harden underwater. Thereafter, hydraulic concrete was used to construct bridges and aqueducts and to line the harbors of Roman ports. Stone and brick masonry became relegated to formwork, the hollow core of which was filled with concrete. The world's largest unreinforced concrete dome (142.5 feet in diameter) still stands on the Pantheon's concrete and brick veneer rotunda walls in Rome.

Major Developments

The Roman method of building with masonry-faced concrete walls, piers, arches, vaults, and domes was lost to the world with the dissolution of the Roman Empire; their technology remained virtually unknown until English inventors began to experiment with making cement in the late 18th century. In 1824, Joseph Aspdin applied for a patent for a process of making cement with English Portland limestone, a name which has become synonymous with cement today.

Builders used a low-moisture concrete placed and compacted with heavy tampers until the development of reinforced concrete in the 1850s necessitated a more wet, fluid mixture. This weaker and less durable poured-in-place concrete was eventually improved by Duff A. Abrams, who developed the water-cement ratio law in 1918. This guideline restricts water content to the lowest value that will allow workability for a particular job. Further experiments in the 1930s proved that vibrating concrete and employing small amounts of well-dispersed entrained air would also eliminate the need for excessive water. Today, the use of stronger, drier cement mixtures is possible due to plasticizers, additives that increase the plasticity or fluidity of concrete without water.

Although not the first to experiment with reinforced concrete, Joseph Monier, a French gardener, is credited with making its use widespread. Monier exhibited his iron-mesh reinforced flower pots and basins at the Paris Exposition of 1867, and subsequently obtained several patents for iron-reinforced concrete basins, pipes, building panels, bridges, and beams.

Monier's display at the Paris Exposition and inspired François Hennébique to develop his own building system using reinforced concrete, which he patented in 1892. Hennébique's Béton Armé system united the column and beam into one monolithic system for construction. Between 1892 and 1902, firms in many parts of Europe licensing Hennebique's patent built several thousand structures with his method.

By the turn of the century, the use of reinforced concrete in buildings was widespread. Experiments with prestressing were also being carried out at the end of the nineteenth century, but none were successful until the engineer Eugene Freyssinet patented a method for long-span bridge design in 1928. Freyssinet used high-strength steel wire to counter the effects of creep in concrete.

Notable Movements, Architects, and Structures

The industrialization of concrete products and structural experimentation characterize 19th-century concrete construction. Relatively inexpensive, standardized concrete building products with codified strengths and characteristics galvanized modernist architects to envision a more egalitarian, socially progressive world. In the 1950s, Architects like Le Corbusier and Oscar Niemeyer chose reinforced concrete as the medium through which to express their socialist ideologies in designing the new capital cities of Chandigarh, India, and Brasilía, Brazil, respectively.

The structural forms of concrete are limited only by human imagination. Auguste Perret insisted in the 1920s that the only form true to reinforced concrete's nature was the trabeated system of monolithic columns and beams organized as frames for infill panels of precast concrete blocks. One of his students, Le Corbusier, proposed that concrete's ability to resist tensile stresses was better illustrated by his Domino system, the cantilevered slab on pilotis. In the search for exploiting concrete's potential, the American engineer Francis S. Onderdork Jr. insisted that the true form of concrete was to be found in the parabolic arch, which efficiently carries its own weight—as demonstrated in London's Royal Horticultural Hall, completed in 1928.

The Greek architect P. A. Michelis argued in the 1950s that the shell structure, developed in the 1930s and 1940s, had exposed the limits of concrete's true potential because the shell utilizes concrete's isotropic nature. A shell structure can carry loads equally in all directions and therefore erase the distinction between load and support. The shell has continued to gain favor as the most innovative and expressive concrete form, as seen in Le Corbusier's Notre Dame du Haut in Ronchamp, France, in Eero Saarinen's TWA Terminal at John F. Kennedy Airport in New York, New York, and most recently in Santiago Calatrava's enormous shell for the Auditorio de Tenerife in the Canary Islands.

Brutalist architecture (the term originates from the French *beton brut*, or raw concrete) brought forth a distinctively different approach to building in concrete. Le Corbusier's Unité d'Habitation of 1952 set the tone for this new style with its angular, massive pilotis, leaving exposed the rough texture of the wood formwork. Other architects working in the Brutalist style approached concrete as if it were a sculptor's material, massive and heavy like stone. Paulo Mendes da Rocha, winner of the 2006 Pritzker Prize, has been honored for his primitive and playful handling of concrete to create buildings such as the Brazilian Sculpture Museum in Sao Paulo. The construction of his massive yet poetic forms in concrete allowed his works to be built quickly and inexpensively.

Recent Trends

Recent technical developments have been in the refinement of casting and finishing techniques, rather than structural experimentation. Advances in chemical formulas for portland cement allow greater control over the qualities desired for different uses, such as high early strength, air-entrained, coloring, and surface finish. Glass-fiber reinforced concrete is a strong light cladding that can be translucent, and fiber optics can be embedded into concrete to transmit light through a wall. Concrete is currently being developed to act as an insulator and to conduct electricity.

Sustainable issues in concrete construction are also of increasing interest. An important trend is the use of waste materials from other industries for cement aggregates and components. The cement industry produces approximately

7 percent of the total greenhouse gas in the world. These emissions can be substantially reduced with the increased use of fly ash (a waste material from electric power plants), instead of portland cement. In addition, high-volume fly ash (HVFA) concrete needs only two-thirds of the water used in conventional concrete. Other methods to enhance concrete's sustainability include minimizing the use of materials for reinforcing and formwork; strategically using concrete's thermal qualities to conserve energy; reducing surface water runoff; and improving concrete quality to prolong building longevity.

Reusable formwork in plastic, metal, and fabric is a more sustainable construction practice and can impart a variety of novel textured finishes to concrete. Another approach recently taken by the architect Tadao Ando is to exploit recycled natural materials. In Awaji Yumebutai, a conference center on an island near Kobe, Japan, Ando used one million scallop shells collected from a local cannery to cover the concrete base of a broad, shallow pool.

END day 1

COMPOSITION OF CONCRETE

Cement

Any adhesive substance capable of uniting non-adhesive materials is called cement. In concrete construction, that substance is portland cement, the most widely used cement in existence.

Portland cement is manufactured from lime, silica, iron oxide, and alumina. The ingredients are properly proportioned, ground, and burned to form clinkers, which are then pulverized to produce cement. This final step is strictly controlled, as the fineness of grinding has a direct bearing on the strength of the cement. Portland cement hardens by reacting to water. The two materials first form a paste that loses its plasticity as it begins to set. The initial set occurs within an hour, and the final set takes about ten hours. The cement continues to harden, however, over a long period of time. Cement paste is the chemically active ingredient in concrete, and is referred to as the matrix.

Table 1.1 summarizes the various types of portland cement used in building construction.

Aggregates

Aggregates are the chemically inert ingredients that are combined with cement and water to make concrete. Aggregates affect the quality of the concrete, reduce shrinkage of the concrete, and serve as a filler, for economy.

Aggregates are classified by size: *fine aggregate* (usually sand) is material ¼ inch or less in diameter, while *coarse aggregate* (gravel or crushed rock) varies in size from ¼ to 1½

Portland Cement	Type of Concrete	Primary Use
Type I	Standard	For general all-purpose use
Type II	Modified	For slow setting and less heat
Type III	High Early Strength	For quick setting and early strength
Type IV	Low Heat	For very slow setting (little heat)
Type V	Sulfate Resisting	For alkaline water and soils

Table 1.1

inches in diameter. For economy, concrete used in massive structures, such as dams, may contain natural stones or rocks ranging up to six inches in size.

The general requirement for natural aggregates is that they should be hard, durable, clean, and free from any harmful matter that would adversely affect the concrete mix. Bank-run gravel aggregates consist of completely rounded shapes that vary in strength, whereas crushed rock aggregates consist of irregular and angular shapes, the strength of which is consistent with the type of rock that is crushed. Irregular or angular-shaped particles are preferred for maximum strength, as water-rounded pebbles may not bond properly with the cement paste. Rounded particles, on the other hand, require less cement paste and improve workability. Most important, the grading of aggregates (the distribution of particle sizes) is critical to a proper mix.

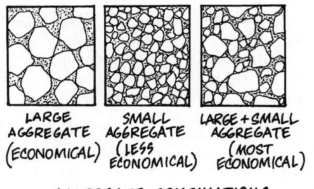

AGGREGATE COMBINATIONS

Figure 1.2

A concrete mix is basically composed of large, coarse aggregate particles between which smaller and finer particles are fitted until all the voids in the mixture are as solidly filled as possible. This requires predetermined percentages of particle sizes, implemented by various sieve tests. Ultimately, when the cement paste is added to a well-graded aggregate mixture,

each particle is coated with the paste and forever bonded to the adjacent particles.

The maximum aggregate size should be no greater than ⅓ the thickness of concrete slabs, or ¾ of the minimum space between reinforcing bars.

Admixtures

Sometimes, materials other than portland cement, water, and aggregates are added to concrete mixes to alter certain characteristics or to achieve some special qualities. These materials are called admixtures, some of which are listed in Table 1.2. In addition to those listed, there are also surface applications or finishes for concrete, which include hardeners, pigments, special aggregates, sealers, abrasives, and fillers for patching. (See later discussion on finishes.) Some portland cements are manufactured with admixtures as part of the cement, such as air-entraining agents, accelerators, and retarders.

CONCRETE MIX DESIGN

Proportioning

The proportioning of the ingredients that comprise concrete is referred to as the *mix*. It involves determining the optimum combination of ingredients to produce concrete that is workable, economical, and has the required strength and other properties when hardened. To satisfy these requirements, several trial mixes may be prepared in which relative amounts of ingredients are varied until the optimum proportions are determined.

The unit of volume for concrete is the cubic foot, and ordinary concrete weighs about 150 pounds per cubic foot. A standard sack of portland cement contains one cubic foot and weighs 94 pounds.

Type of Admixture	Ingredients	Principal Use
Accelerators	Calcium chloride	Speed up setting time
Air-entraining agents	Resins, fats, and oils	Resist freezing action
Retarders	Starches, sugars, and acids	Slow down setting time
Waterproofing	Stearate compounds	Decrease permeability
Water-reducing	Organic compounds	Reduce water content
Workability agents	Powdered silicas and lime	Improve workability

Table 1.2

Aggregates are measured either by weight or volume (dry sand and gravel weigh about 100 pounds per cubic foot), and water is measured by the gallon, which weighs about 8 pounds.

For small, relatively uncontrolled concrete work, the mix is generally expressed by volume. For example, a 1:3:5 mix consists of one part cement; three parts fine aggregate (sand); and five parts coarse aggregate (crushed rock or bank-run gravel).

Water-Cement Ratio

A number of factors are important in concrete mix design, including the grading and proportioning of aggregates. But perhaps most important is the water-cement ratio. This ratio, expressed as the number of gallons of water for each sack of cement, is the major factor controlling concrete strength, and to a large extent, its durability. Maximum strength is obtained by using the minimum amount of water required to complete hydration of the cement, but a mix of this type would be too dry and unworkable. Thus, a plastic or workable mix always contains more water than the amount needed to attain maximum strength. In other words, concrete strength decreases as the extra water required for workability increases.

The optimum water-cement ratio provides the minimum amount of cement paste that will coat each aggregate particle and fill all voids while providing the required concrete strength and adequate workability. Excess water not only reduces the concrete's strength and durability, but it may also produce *laitance*, which is a chalky surface deposit of low strength. If laitance does appear, it must be removed before any new concrete is poured, in order for the new concrete to bond to the old.

Strength

The compressive strength of concrete always refers to its strength 28 days after being placed. Concrete develops a wide range of compressive strength values, depending primarily on the water-cement ratio. For example, a mix using about seven gallons of water for each sack of cement will develop a compressive strength of about 3,000 psi after 28 days. While it is possible to obtain concrete with a compressive strength of 10,000 psi or greater, the usual range for structural concrete is 3,000 to 6,000 psi, with values around 4,000 psi being the most common. High-early-strength concrete can develop a compressive strength in 7 to 14 days equal to that developed by normal concrete in 28 days.

MIXING

All concrete should be thoroughly mixed until it is uniform in appearance and all ingredients are evenly distributed. For small jobs, concrete may be mixed at the job site, but this method is slow, relatively imprecise, and generally not recommended for quantities in excess of a few cubic yards. More commonly, concrete is mixed in a plant, a truck, or a combination of both.

Ready-mixed concrete is mixed completely at a central mixing plant and transported to the building site in an agitator truck with a revolving chamber. Ordinarily this type of concrete must be placed within one and a half hours after water is added to the mix. Therefore, jobs using ready-mixed concrete are limited to those relatively close to the mixing plant.

Transit-mixed concrete is mixed completely in a truck mixer. The dry materials are picked up at a central plant and placed in a mixer truck that carries a water tank. The mixer continues to revolve the dry mix while en route to the site; after arrival, water is added and mixed, and the concrete is deposited.

The transit-mix truck may also pick up the concrete at the plant in a partially mixed state and complete the mixing in the truck.

FORMWORK

Forms are the molds into which the concrete is placed and held in shape until it has hardened and developed sufficient strength to support its own weight. Formwork may be job-constructed or prefabricated units of standard lumber, plywood, metal, fiberboard, paper pulp, or a variety of reinforced synthetics.

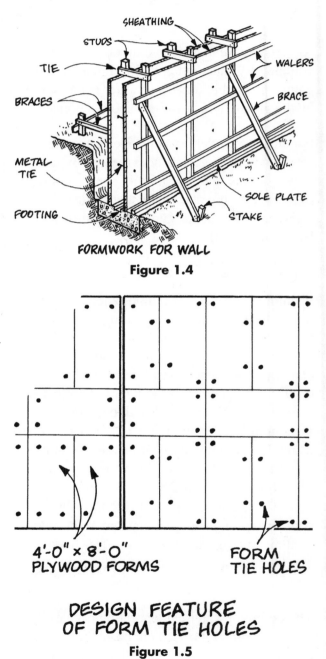

FORMWORK FOR WALL

Figure 1.4

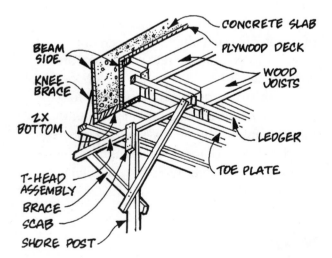

SLAB AND BEAM FORM
(VIEWED FROM BELOW)

Figure 1.3

DESIGN FEATURE OF FORM TIE HOLES

Figure 1.5

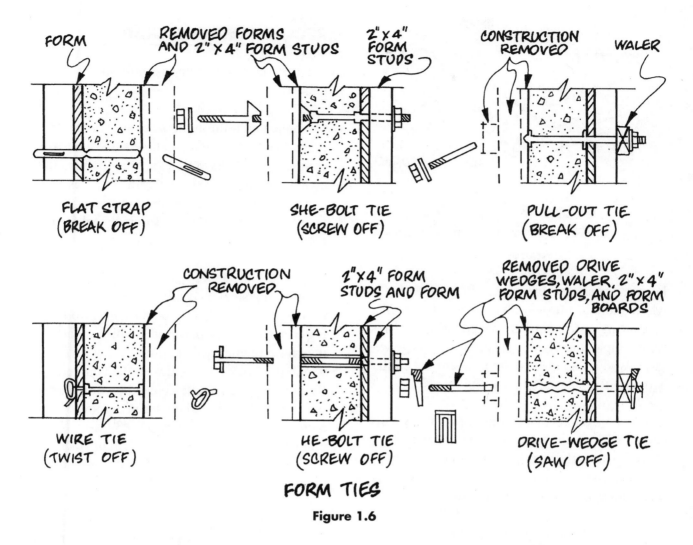

FORM TIES

Figure 1.6

The most important requirement for all formwork is that it be strong and stiff enough to support the considerable weight and fluid pressure of wet concrete. Forms must also be tight so that there will be no significant loss of cement paste. Finally, as the plastic concrete will take on the shape, texture, and peculiarities of the forming material, the choice of formwork should be carefully considered from the standpoint of aesthetics.

Form ties are metal devices used to prevent concrete forms from spreading. The location of form ties is an important design factor in wall surface appearance, as shown in Figure 1.6.

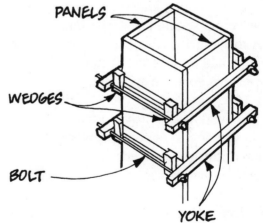

COLUMN FORM

Figure 1.7

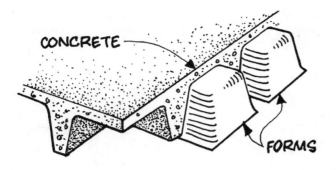

STEEL FORMS FOR WAFFLE PATTERN CEILING
Figure 1.8

Forms, or parts of forms, are often omitted where there is a firm earth surface capable of supporting or molding the concrete. In footings, for example, the bottom of the footing is generally cast directly against the undisturbed earth and only the sides are molded in forms.

Before concrete is placed, the forms are usually coated with a suitable oil or other material to prevent water absorption or bond between the form and the concrete. Excessive oiling should be avoided, however, as it may permanently stain the concrete or otherwise interfere with the final concrete finish. It is important that the oil or other coating material be applied to the forms before setting the steel reinforcement in order to avoid accidental coating of the steel, which would prevent satisfactory bonding between the concrete and steel.

All formwork is expensive, often representing the major cost of concrete work. For this reason, forms should be designed and constructed so that they are practical to erect, simple to strip, and capable of being reused. In many areas, the formwork for large buildings is designed by a structural engineer.

end day 2

REINFORCED CONCRETE

Concrete is strong in compression but weak in tension.

Concrete and reinforcing steel are compatible materials: they have almost the same coefficient of thermal expansion so that temperature changes do not introduce significant stresses, concrete is sufficiently impervious and fire-resistant to protect the reinforcing steel from corrosion and fire, and concrete bonds well to steel.

Reinforcing steel is usually in the form of round bars or welded wire fabric. Reinforcing bars (rebar) are deformed, which means they have regular deformations, or projections, which interlock and bond with the concrete. Reinforcing bars are designated by numbers representing the bar diameter in eighths of an inch. The standard sizes are #3 (⅜-inch diameter) through #11 (1⅜-inch diameter); #14 (1¾-inch diameter); and #18 (2½-inch diameter).

Reinforcing bars are specified by their ASTM designation, such as ASTM A615 (billet steel); ASTM A616 (rail steel); A617 (axle steel); or ASTM A706 (low-alloy steel). In addition, a grade designation is used: grades 40, 50, 60, and 75 have minimum specified yield strengths of 40,000, 50,000, 60,000, and 75,000 psi, respectively. The most common designation for reinforcing bars used in buildings is ASTM A615 Grade 60.

Reinforcing bars are furnished with rolled-in markings that identify the rolling mill, the bar size number, the type of steel, and an additional marking for identifying the higher strength steels, as shown in Figure 1.9.

NUMBER SYSTEM   LINE SYSTEM

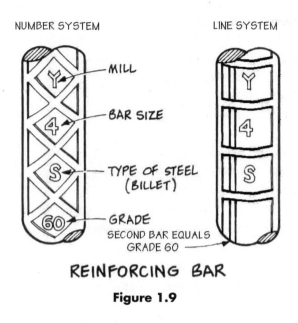

MILL

BAR SIZE

TYPE OF STEEL
(BILLET)

GRADE
SECOND BAR EQUALS
GRADE 60

REINFORCING BAR

Figure 1.9

Welded wire fabric (WWF) is a grid of smooth or deformed cold-drawn steel wires welded at all points of intersection. This type of reinforcement may be used for floors, walls, roofs, or other large expanses of concrete.

Welded wire fabric is designated by the size and spacing of the longitudinal and transverse wires. For example, WWF 6 × 6 – W2.9 × W2.9 indicates welded wire fabric with wires at six inches on center in each direction, with each wire having a cross-sectional area of 2.9 ÷ 100 or 0.029 square inches (which is six-gauge wire).

Steel used in prestressed concrete will be discussed later in this lesson.

Steel reinforcement must be clean, accurately placed, and have a sufficient protective covering of concrete. Before placement, reinforcing steel should be cleaned of all loose rust, oil, mud, paint, or any other foreign matter that might reduce the bond between the bars and the concrete.

Reinforcing bars are generally spliced by lapping them a specified number of bar diameters.

In some cases, bars may be butted end-to-end and spliced by welding or using mechanical devices.

Reinforcing bars may be pre-assembled. Column reinforcement, for example, is often completely assembled as a cage before being placed within the formwork. The proper placement of reinforcement is shown on the structural drawings, which are then used as the basis for preparing shop drawings. When concrete is placed over a network of steel reinforcement, the pressures may tend to dislodge the steel. Therefore, the reinforcement must be held rigidly in place, with sufficient supports and ties, so that the steel will end up in precisely the right location. For this purpose there are a number of devices, such as ties, chairs, and supports, that hold the reinforcing steel in its exact location.

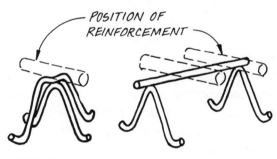

POSITION OF
REINFORCEMENT

HIGH CHAIR   CONTINUOUS HIGH CHAIR

SLAB BOLSTER   BEAM BOLSTER

REBAR SUPPORT DEVICES

Figure 1.10

Steel reinforcement must have adequate concrete coverage to protect it from corrosion. For example, footings should have at least three inches of concrete between the steel and the ground. Retaining walls require two inches of concrete cover; beams and columns, 1½ inches; and slabs, ¾ inch.

Where concrete structures are exposed to chlorides, such as deicing salts or seawater, reinforcing bars may be epoxy-coated or galvanized to resist corrosion.

LIGHTWEIGHT CONCRETE

Structural Lightweight Concrete

One of the principal disadvantages of concrete construction is its great weight. By using lightweight aggregates made from expanded shale or clay instead of natural aggregates, structural lightweight concrete can be produced. Its weight of 90 to 115 pounds per cubic foot is substantially less than that of normal concrete, which weighs 150 pounds per cubic foot.

Structural lightweight concrete with strength, workability, and appearance comparable to normal concrete can usually be obtained. There are some differences, however.

For structural lightweight concrete, the maximum size of coarse aggregate is generally ¾ inch, and air entrainment is almost always used. Handling and placing are easier; the modulus of elasticity is lower (and thus deflections are greater); drying shrinkage is slightly greater; thermal insulation properties are better, and the cost is almost always greater than that of normal-weight concrete.

Insulating Lightweight Concrete

Another type of lightweight concrete is insulating lightweight concrete, used primarily for thermal insulation in roof construction. It has a weight of 15 to 90 pounds per cubic foot and

relatively low compressive strength. It is usually made with aggregates of expanded materials such as perlite or vermiculite, although other lightweight aggregates may be used. Some mixtures are made without fine aggregate (no-fines), which creates voids in the concrete.

Another way of producing insulating lightweight concrete is to incorporate a uniform cellular structure of air voids into a cement paste or cement-sand mortar by using preformed or formed-in-place foam.

PLACEMENT OF CONCRETE

Prior to placing concrete, trenches and formwork should be thoroughly cleaned out, reinforcement should be checked for position, and wood forms should be moistened. Concrete may be placed by spouts or chutes; pumped; placed pneumatically (gunite); or even deposited under water with the use of a tremie. In any event, concrete must be placed evenly, continuously, and—most important—in a manner that avoids segregation of the aggregates.

When new concrete is placed on hardened concrete, precautions must be taken to ensure a well-bonded, watertight joint. The surface of the hardened concrete should be thoroughly moistened and prepared using sandblasting, water jets, pneumatic tools, or other methods.

Concrete should be placed as close as possible to its final location. A load of concrete should not be dumped at one point and permitted to flow over long distances, as this will lead to the segregation of water and fine particles from the rest of the mass.

Dropping the mixture from excessive heights will also cause it to separate. Vertical drops should generally be limited to four feet.

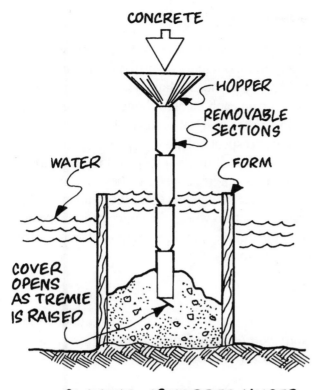

PLACING CONCRETE UNDER WATER WITH A TREMIE
Figure 1.11

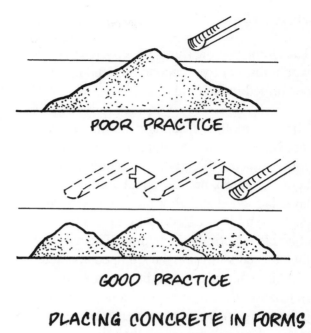

PLACING CONCRETE IN FORMS
Figure 1.12

When concrete is placed, air bubbles are frequently trapped in the body of the mixture. If allowed to remain, these air bubbles will produce a honeycomb effect that can cause a substantial reduction in concrete strength and watertightness, as well as an unsightly finish. To eliminate this condition, concrete should be compacted or consolidated by hand tools or by the use of mechanical vibrators. The form-work can be vibrated externally, or vibrators can be immersed in the mix itself. Vibration of concrete results in greater density, homogeneity, durability, and more complete contact with the reinforcing. Vibration also allows the use of stiffer mixes with a reduced cement content. However, vibration does not make the concrete stronger, and excessive vibration may result in segregation.

Concrete may be placed during almost any weather condition, but special methods and procedures are required during both extremes of hot and cold weather. Concrete must not be allowed to dry out too rapidly or to freeze.

Pneumatically applied concrete, often called gunite, is shot into place using compressed air. It may be used for repair work as well as new construction. It is especially useful for placing concrete that has a large surface area and thin sections, such as the sides and bottoms of swimming pools and thin sections used to strengthen existing masonry walls for seismic forces.

TESTING

In order to verify that the concrete used on a project has the required quality, several tests may be performed. Among these are the *slump test* and the *cylinder test*.

Slump Test

The slump test measures the consistency and workability of the concrete mix and is usually performed in the field.

The test employs a standard slump cone mold that is set on a level surface and filled with concrete taken directly from the mixer, in three layers of approximately equal volume. Each layer is rodded and after the top layer is leveled, the mold is carefully lifted and the wet concrete is permitted to slump. The mold is then placed alongside the slumped concrete and the difference in height measured directly. In general, the acceptable slump for concrete in building construction ranges from two to six inches. If the slump is less than two inches, the mix is probably too stiff and unworkable, and if it is greater than six inches, it may be too wet and loose.

After the measurement is completed, the slumped concrete may be tapped gently, which will give some indication of its workability.

Cylinder Test

The cylinder test measures the compressive strength of concrete and utilizes standard test cylinders 6 inches in diameter and 12 inches long. From each batch of concrete, at least two cylinders are cast, laboratory-cured for 7 and 28 days, respectively, and tested in a crushing machine. If the cylinders fail to attain the specified strength, cores 2 inches to 4 inches in diameter may be cut from those areas in the actual structure where the concrete is suspected of being deficient and tested in a laboratory for compression.

Other Tests

Other tests that may be performed include the *Kelly ball test*, the *impact hammer test*, and *tests for air content*.

The Kelly ball test, like the slump test, measures workability. In this test, a 30-pound, six-inch diameter hemisphere is dropped from a standard height onto the surface of the fresh concrete, and its penetration into the concrete is measured. When properly calibrated, the results

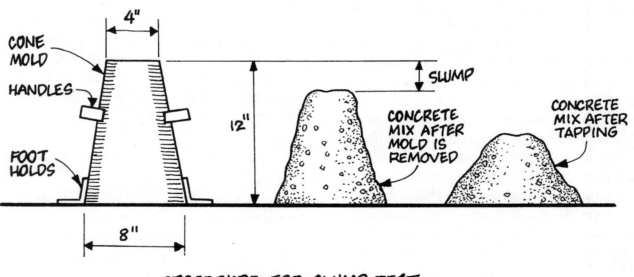

PROCEDURE FOR SLUMP TEST

Figure 1.13

of the Kelly ball test can be directly related to slump.

The impact hammer test is a common nonde-structural test, in which the rebound of a spring-loaded plunger is measured after it strikes a smooth concrete surface. The rebound reading gives an approximate indication of the concrete strength. This test is simple and nondestructive, but it is not a substitute for standard cylinder tests.

There are various methods for determining air entrainment in concrete that measure the volume of air contained in the mix. These tests should be regularly made immediately after the concrete is discharged from the mixer, and frequently after the concrete has been placed and consolidated.

END DAY 3

CURING

The curing of concrete consists of maintaining the proper humidity and temperature for some period of time after it is placed, to assure satisfactory hydration of the cement. Excessive evaporation of water can retard the hydration process and reduce the strength of the concrete, as well as causing the concrete to shrink and develop surface cracks. Exposed surfaces must therefore be protected against moisture evaporation.

Concrete may be cured by the following methods:

1. Supplying additional moisture to the concrete surface, by ponding or sprinkling
2. Using a wet covering, such as moist sand, burlap, or straw
3. Covering the surface with a membrane or curing compound that prevents evaporation
4. Leaving wood forms in place and keeping them moist

The length of the curing period varies from 3 to 14 days or more, depending on many factors. As all the desirable qualities of concrete are improved by curing, the curing period should be as long as possible, consistent with other project requirements.

The most favorable temperature for curing concrete is between 50 and 70°F. The chemical reaction of hydration generates internal heat in the concrete, which must be considered,

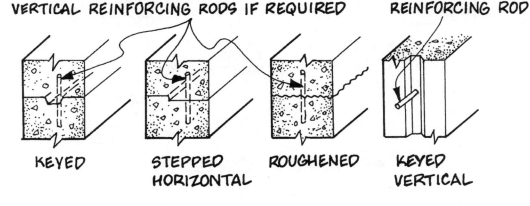

VERTICAL REINFORCING RODS IF REQUIRED   REINFORCING ROD

KEYED   STEPPED HORIZONTAL   ROUGHENED   KEYED VERTICAL

CONCRETE CONSTRUCTION JOINTS

Figure 1.14

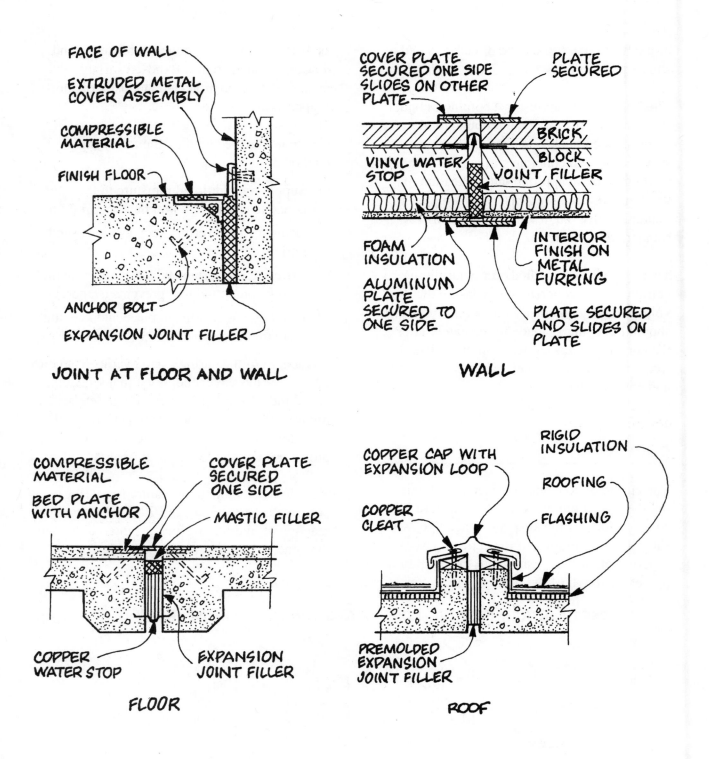

FACE OF WALL
EXTRUDED METAL COVER ASSEMBLY
COMPRESSIBLE MATERIAL
FINISH FLOOR
ANCHOR BOLT
EXPANSION JOINT FILLER

JOINT AT FLOOR AND WALL

COVER PLATE SECURED ONE SIDE SLIDES ON OTHER PLATE
PLATE SECURED
BRICK
BLOCK
VINYL WATER STOP
JOINT FILLER
FOAM INSULATION
ALUMINUM PLATE SECURED TO ONE SIDE
INTERIOR FINISH ON METAL FURRING
PLATE SECURED AND SLIDES ON PLATE

WALL

COMPRESSIBLE MATERIAL
BED PLATE WITH ANCHOR
COVER PLATE SECURED ONE SIDE
MASTIC FILLER
COPPER WATER STOP
EXPANSION JOINT FILLER

FLOOR

COPPER CAP WITH EXPANSION LOOP
RIGID INSULATION
ROOFING
FLASHING
COPPER CLEAT
PREMOLDED EXPANSION JOINT FILLER

ROOF

EXPANSION JOINTS

Figure 1.15

especially in hot weather. In very hot weather special care may have to be taken, such as cooling the concrete ingredients or adding ice to the mixing water. Wind and sun tend to dry out concrete, and temporary windbreaks and/or shading may be required.

During cold weather, additional heat may be required from heaters, heating coils, or live steam. It may even be necessary to temporarily enclose and heat portions of the building during and after concrete placement.

CONCRETE JOINTS

Construction Joints

The horizontal or vertical joints between two successive concrete pours are called construction joints. These joints exist whenever concreting is interrupted, or when new concrete is placed against old. Joints should be straight, and their form and location should be designed carefully.

To ensure the maximum bond between old and new pours, the old surface should be roughened, cleaned, and wetted before placing the new concrete. In many cases, reinforcing bars extend through the joint and help to connect the new and previously poured concrete. Construction joints are always planes of weakness, and therefore should be located at sections of minimum shear. They are often keyed to provide some shear strength.

Expansion Joints

Expansion joints are designed to allow free movement of adjacent parts due to expansion or contraction of the concrete. This movement may be caused by shrinkage or changes in temperature. Expansion joints provide complete separation through a structure, from the top of

the footings to the roof. They are waterproof, weather tight, and generally filled with an elastic joint filler. The placement and size of joints is a function of the size of the building and the maximum expected temperature differential.

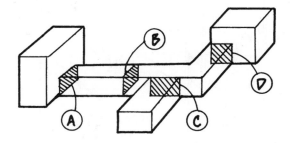

(A) LOW BLDG. ABUTTING HIGH BLDG.

(B) CONCRETE BLDG. OVER 200' LONG

(C) INTERSECTION OF BLDG. WINGS

(D) NEW BLDG. ADJOINING EXISTING BLDG.

EXPANSION JOINTS
Figure 1.16

Expansion joints are required in buildings more than 200 feet long, at joints of building wings, and at additions to existing buildings.

Control Joints

Control joints are tooled, sawed, or premolded joints to allow for the shrinkage of large concrete areas. Control joints create a weakened section that induces cracking to occur along the joint, rather than in a random fashion.

Isolation Joints

Isolation joints provide a separation between a slab on grade and columns or walls, so that each can move independently.

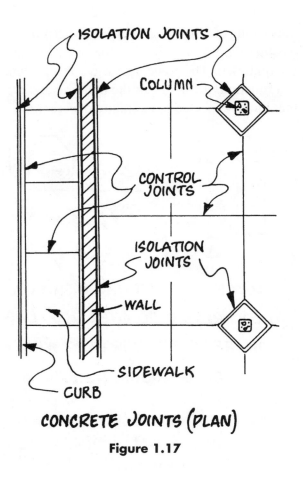

CONCRETE JOINTS (PLAN)

Figure 1.17

PRESTRESSED CONCRETE

Prestressed concrete is concrete placed in compression by applying a tensile force to prestressing steel *before* the external loads are applied. The purpose is to cause stresses in the concrete that are opposite in direction from those caused by the external loads.

The combined effect of the prestress and the external dead and live loads usually results in compression over the entire cross-section of the beam, in contrast to the relatively small compression area of conventional reinforced concrete beams. Prestressing thus results in more efficient and economical use of material, especially in repetitive long-span applications. Concrete members can be smaller, span greater distances, and support greater loads.

Because prestressed members are completely in compression, tension cracks are prevented, which is particularly advantageous in structures exposed to the weather. In addition, prestressed concrete members are stiffer (because the entire section is effective) and have greater shear strength.

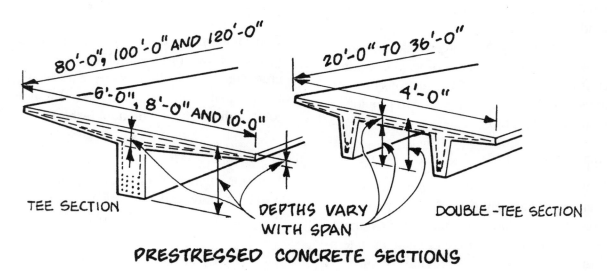

PRESTRESSED CONCRETE SECTIONS

Figure 1.18

However, offsetting these advantages are greater material and labor costs and the need for closer quality control than with conventional reinforced concrete.

There are two procedures used to apply prestress: *pretensioning* and *posttensioning*.

In pretensioning, the prestressing steel is stretched between abutments, usually at a casting yard away from the building site, and tensioned by jacks *before* placing the concrete. After the concrete is placed and has attained its required strength, the prestressing steel is cut, and its tensile force is applied as a compressive force to the concrete through bond between the concrete and steel.

Some loss of prestress occurs because of creep and shrinkage, slip, and friction.

Pretensioning lends itself to mass production, as the casting beds can be hundreds of feet long, the entire length can be cast at one time, and individual beams can be cut to the required lengths.

In a posttensioned beam, the concrete is cast with a hollow duct or sleeve to encase the prestressing steel and prevent bond between the concrete and steel. This is usually done on the site. The concrete is cured and after it has acquired sufficient strength, the steel is tensioned by jacking against anchorages at the ends of the beam, which compresses the concrete. The prestressing steel is permanently locked under stress by special end anchors and grouted. The losses after tensioning, caused by friction, elastic shortening, and shrinkage, are usually less than with pretensioning.

Both the concrete and steel used in prestressed concrete construction must have greater strength than required for conventional rein-

forced concrete. For example, the concrete is generally required to have a compressive strength of 5,000 psi or greater.

There are three types of prestressing steel: high-strength bars, single wires, and wire strands, all of which are called tendons. Wire strands have seven wires each: one central wire enclosed tightly by six helically placed outer wires.

END day 4

PRECAST CONCRETE

Precasting is the casting of concrete members at a location other than its final position in the structure, as opposed to cast-in-place concrete, which is produced in the exact location it will occupy. Precasting can be done almost anywhere, unless the handling and transportation of members become unfeasible. Thus, it is good practice to verify the manufacturing location, the size of precast elements, trucking sizes, and highway clearances relative to the site location.

Precast concrete has the greatest economic advantage when there are many identical members to be cast, as the same forms can be reused many times. There are other advantages to precasting:

1. Better quality control of concrete
2. Better control over curing (may be steam cured)
3. Members castable in all weather
4. Members erectable in all weather
5. Faster actual construction time

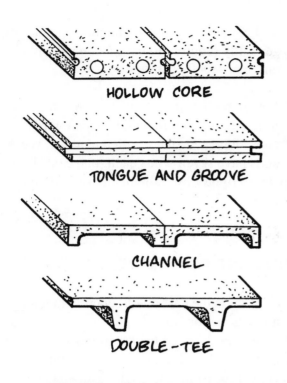

PRECAST CONCRETE PLANKS

Figure 1.19

Precast Floor and Roof Systems

Precast floor and roof systems, which are often prestressed, consist of a variety of precast planks that may be used with precast beams, joists, and purlins. Curtain wall panels may also be precast. Once on the job, precast members are handled in much the same way as other building components as they are integrated with the structure.

Tilt-Up Construction

Tilt-up construction involves casting a wall panel in a horizontal position and then tilting it to its final vertical position. Usually, these panels are solid, reinforced slabs five to eight inches thick and long enough to span between the supporting columns or footings. Various casting platforms may be used, the most common of which is the slab on grade, which serves as the bottom form. It is important to prevent bonding of one surface to another; therefore, bond-breaking agents, patented liquid solutions, and various sheet materials are used between the slab and the tilt-up panel.

The upper or exterior surface of the panel may be finished by floating, troweling, or with special decorative finishes, such as pebbled, grooved, polished, colored, etc. Tilt-up panels may be load- or non-load-bearing, may rest on a foundation wall or span between footings, and are connected to cast-in-place columns.

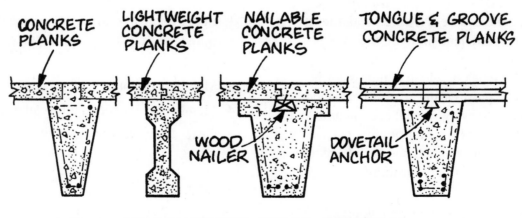

PRECAST CONCRETE JOIST SYSTEMS

Figure 1.20

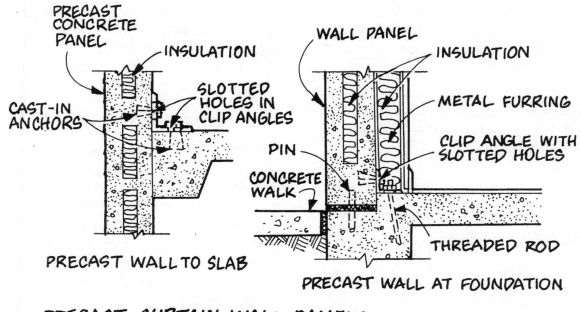

PRECAST CURTAIN WALL PANELS

Figure 1.21

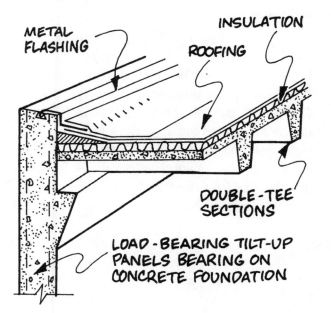

TILT-UP PANEL SYSTEMS

Figure 1.22

Lift-Slab Construction

Lift-slab construction consists of casting floor and roof slabs of a multistory building, one upon another, with a membrane or bond-break-ing agent between, to prevent bonding of the separate slabs. Jacks attached to the structure's columns lift the slabs to their final position, where they are welded into place using special steel collars. This method of construction eliminates practically all formwork, and greatly shortens the overall construction time.

A great advantage of lift-slab construction is that all mechanical pipes, conduits, and ducts can be installed on grade. Thus when the floors are raised into their permanent positions, all electrical, heating, plumbing, ventilating, and air-conditioning have been completely roughed in.

Tube-Slab

Tube-slab refers to a system in which paper tube fillers are embedded in the section to obtain a flat ceiling with no exposed beams. This allows mechanical and duct spaces to be integrated within the thickness of the system.

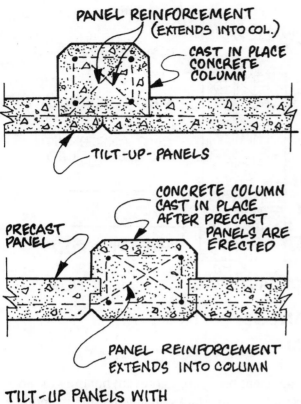

PANEL REINFORCEMENT
(EXTENDS INTO COL.)

CAST IN PLACE
CONCRETE
COLUMN

TILT-UP-PANELS

PRECAST
PANEL

CONCRETE COLUMN
CAST IN PLACE
AFTER PRECAST
PANELS ARE
ERECTED

PANEL REINFORCEMENT
EXTENDS INTO COLUMN

TILT-UP PANELS WITH
CAST-IN-PLACE COLUMNS
Figure 1.23

CONCRETE FINISHES

Concrete finishing is performed in many different ways, depending on the final surface effect desired. In general, finishes are limited only by the designer's imagination and the properties of the concrete. Finishes for walls and ceilings are generally applied to concrete that has already set, while finishes on floor slabs are applied while the concrete is still plastic and workable.

Walls and Ceilings

Walls can be finished in a variety of ways: rough; smooth; with a rubbed finish; sandblasted; with textured formwork; bush hammered; with exposed aggregate; or with an applied finish such as stucco, plaster, ceramic tile, or concrete paint. They may also be furred and insulated and then covered with a variety of materials such as plasterboard, composition panels, and plywood.

Ceilings may be finished in many of the same ways as walls, in addition to adhesive-applied acoustical tile.

Floors

Floor surfaces may be finished with a wood float finish; a smoother and harder steel troweled finish; an applied texture, such as brooming; or an applied finish, such as nonslip, pigmented, heavy duty, etc. The finishing operation of all slabs involves screeding or leveling, floating, and troweling; and at any point in the process a final finish can be produced. The final

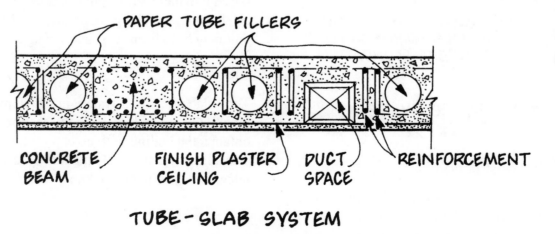

PAPER TUBE FILLERS

CONCRETE
BEAM

FINISH PLASTER
CEILING

DUCT
SPACE

REINFORCEMENT

TUBE-SLAB SYSTEM
Figure 1.24

finishing of a slab, however, is not done until all surface water has disappeared.

Terrazzo

Terrazzo is not, strictly speaking, a concrete finish, but rather a topping material that is applied over concrete slabs. Terrazzo is a mixture of portland cement and water, known as the matrix, to which colored marble granules are added. After the mixture has set, it is ground and polished to a smooth finish. Terrazzo can be applied to floors, walls, wainscots, and stairs. It is also precast as stair treads, window stools, saddles, and floor tiles.

end day 5

Seamless terrazzo floors and precast terrazzo tile use a synthetic resin matrix instead of cement and water. Further information can be found in Lesson 7, under Flooring.

LESSON 1 QUIZ

1. If you are pouring a large concrete slab and wish to avoid random cracks caused by shrinkage, you would likely provide

   A. expansion joints.

   B. isolation joints.

   C. control joints.

   D. construction joints.

2. When pouring concrete in hot weather, what kind of agent would one add to the mixture?

   A. Retarding agent

   B. Air-entraining agent

   C. Hardening agent

   D. Workability agent

3. The purpose of a concrete slump test is to

   A. determine the plasticity of the mix.

   B. measure the consistency and workability of the mix.

   C. test the compression of the mix.

   D. define the compression strength of the mix.

4. Which of the following qualities are normally associated with lightweight concrete?

   I.   Good thermal insulation

   II. High density

   III. Excellent fire resistance

   IV. Low cost

   V.   Difficulty of handling

   A. I and III   C. I, III, and V

   B. II and III   D. II, IV, and V

5. The principal determinant of concrete strength is the

   A. amount of air entrained.

   B. type of portland cement used.

   C. volume of aggregate in the mix.

   D. water-cement ratio.

6. In curing concrete slabs, which of the following procedures is rarely used?

   A. Overlay with plastic membrane

   B. Spray with moisture

   C. Blow with fans

   D. Cover with sand

7. While observing a concrete pour, an architect notices that the entire concrete load has been deposited from a 6-foot height at the middle of a 40-foot-long form. The architect should direct the contractor to

   A. spread the concrete throughout the length of the form.

   B. pour the next two concrete loads at each end of the form.

   C. pour the next load in exactly the same location.

   D. remove and discard the concrete.

8. The steel in reinforced concrete

   A. furnishes rigidity.

   B. provides tensile strength.

   C. increases bond strength.

   D. adds ductility and durability.

9. The size of coarse aggregate in a concrete mix is governed by the

    I. type of cement used.

    II. strength of concrete desired.

    III. thickness of concrete section.

    IV. water-cement ratio.

    V. space between reinforcing bars.

    A. V only   **C.** II and III

    B. I and IV   **D.** III and V

10. All of the following are advantages of using precast concrete sections *EXCEPT*

    A. better quality control.

    B. faster curing.

    C. all-weather construction.

    D. greater economy.

MASONRY

INTRODUCTION

Masonry construction consists of walls and other building elements made of individual masonry units that are joined together with mortar. Small masonry units are either solid or hollow and include brick, structural clay tile, concrete block, gypsum block, and glass block. Larger units include terra cotta, ceramic veneer, precast concrete, and various types of stone.

HISTORY

Discovery and Early Uses

The dry stacking of shaped stones to form walls as boundaries and shelter dates back to the earliest human cultures in the Paleolithic era, otherwise known as the Stone Age. However, the quarrying and cutting of stone was not practiced regularly until bronze tools were developed in the West around the third millennium BCE. At this time, the Egyptians began to build pyramids as tombs constructed of precisely cut and dressed blocks of limestone, basalt, and granite weighing from two to four tons each.

The earliest bricks of uniform shape date to 7500 BCE, found in Neolithic settlements in southeastern Turkey. From the Middle East to the Indus valley, sun-dried mud brick was made in the proportions 4:2:1, which is still considered the best ratio for effective bonding. The next technological development was the transition from sun-drying to firing in a kiln for improved durability. By 4000 BCE, the people of Mesopotamia were building palaces and ziggurats of sun-dried and fired mud brick. The ancient Roman military introduced fired brick to many parts of its empire through the use of mobile kilns.

Corbelled arches and vaults provided a fire-proof cover for buildings throughout the ancient world; however, they could only cover limited spans. The ancient Romans discovered an ingenious remedy to this problem. The "true arch," constructed of *voussoirs* wedged in compression, could span tremendous distances. This exploitation of masonry's compressive strength made way for the later developments of the barrel vault, groin vault, and hemispherical dome. With the invention of concrete, masonry was primarily used as the exterior formwork of walls and vaulted structures. However, the masonry true arch was still used extensively for the construction of bridges and aqueducts. Medieval builders of the Romanesque period attempted to emulate ancient Roman vaulted structures, but were limited to the use of stone. With the knowledge of concrete lost, they were only able to build structures of a limited span.

Major Developments

The structural possibilities of stone and brick masonry were advanced with the building of Gothic cathedrals in Western Europe. Gothic masons discovered that they could construct masonry vaults at greater heights and width with the addition of a system of flying buttresses designed to absorb the excess weight not taken in by the structure's piers. In fact, with a properly designed pier and buttress system, the walls between the piers were not needed to defer any load other than their own

weight, therefore making possible the inclusion of large expanses of stained glass between structural members. The competition between cities in the 13th century drove medieval masons to experiment with increasing heights. Builders of Beauvais Cathedral reached the technological limit, with vaulting in the choir reaching 157.5 feet. Twelve years after its completion, the vault collapsed, probably due to high winds.

Although ashlar stone masonry continued to be the chosen material of the wealthy, brick declined in status. During the Renaissance and Baroque periods, brick walls were seen as crude and were often covered with a thin veneer of plaster to give them the appearance of more costly stone. Brick rose in status again after the Industrial Revolution of the late eighteenth century. Machines replaced the manual labor involved in molding and firing bricks, leading to a greater uniformity of brick size, weight, strength, and color. Mechanized methods for quarrying and cutting stone reduced costs, and brought brick and stone in closer price range. The development of portland cement mortar in the 19th century brought even greater strength to masonry buildings.

In the early 19th century, the brick cavity wall was invented to resist water penetration. A two-inch air space was left between two masonry walls to keep the interior wall dry. The dual walls also provided greater support for the heavy loads of upper floors and the roof. In the 1850s, a special brick with air holes was developed. Around the same time, the reinforced brick wall came about with the development of reinforced concrete. The cavity space between the brick walls was filled with portland cement and metal reinforcing.

Prismatic glass blocks were developed in the early 19th century to provide natural light for masonry industrial factories. Greater uses for glass block were found after the invention of reinforced concrete and the mechanized manufacturing of glass in the early 20th century. In 1907, the German engineer Friedrich Keppler applied for a patent for interlocking solid glass blocks designed to be used as masonry walls within reinforced concrete structures. It was not until the 1930s that hollow glass block such as that used today was developed by the Owens Illinois Glass Corporation.

Although the ancient Romans used a large, solid concrete brick for construction in the region of Naples, hollow concrete block such as that used today is a recent invention. In 1900, Harmon S. Palmer applied for a patent for a cast iron machine that could mass-produce hollow blocks. However, the blocks were so heavy that they had to be lifted into place with a crane. By 1905 thousands of American companies were manufacturing this block; a few years later the machine was made available for purchase from the Sears and Roebuck catalog. In 1917, F. J. Straub's new cinder block solved the excessive weight problem, and by the early 1930s cinder block construction was widespread. Today, lighter weight block is made by replacing the sand and gravel with clay, shale, slate, or plastic beads.

Notable Movements, Architects, and Structures

Until the late 19th century, stone and brick masonry were the most widely used building material. The stately appearance of Henry Hobbs Richardson's Neo-Romanesque buildings in Boston could only achieve their grandeur in heavily rusticated stone. However, the development of concrete, cast iron, wrought iron, and later steel, quickly replaced masonry as the major structural systems in multistory buildings. Masonry today is mostly used for infill panels, cladding, partitions, and flooring.

Around the turn of the 20th century, the American architect Louis Sullivan gave brick a new status by showcasing its ornamental beauty with the enormous brick arches of the 1908 National Farmer's Bank in Owatonna, Minnesota. Frank Lloyd Wright used long, flat Roman brick and cut limestone copings to emphasize the long horizontal lines of prairie houses such as the Robie House, built in Chicago in 1906. Modern architects of the twentieth century emphasized the thin, flat, horizontal quality of the brick wall, seeing the arch as anathema, a historical quotation. The underlying purpose of brick was questioned by the modern architect Louis Kahn in the early 1970s, when he asked: "What do you want, Brick?" The irrefutable answer was: "I like an Arch." And thus, Kahn brought the arch back into the modernist vocabulary.

Wright also experimented with making his own hand-crafted "textile" concrete blocks to build a series of California houses, the most well-known being the Hollyhock House in Los Angeles of 1924. Perhaps the earliest example of glass block in residential architecture was the 1931 Maison de Verre in Paris. This collaboration between Pierre Chareau and Bernard Bijvoet for a doctor was, at the time, a shockingly modern exposé of steel, glass and glass block. In order to demonstrate glass block potential in construction, a 108-foot high tower constructed of glass block and blue plate glass was built for the Glass Center the New York World's Fair in 1939, an exhibition hall to showcase American glass products.

Recent Trends

The trend in brick masonry today is toward a new variety of designer colors and textures and the production of non-standard sizes. Cast stone is now available in colors to match. Prefabricated masonry walls and new mortars and adhesives are making it easier to place masonry. The mortar-free, drystack system is the latest development for concrete clock and brick. Glass blocks are now available in a variety of colors and textures, or can be illuminated with fiber optics.

In terms of sustainability, stone, brick, and concrete masonry are fairly earth-friendly products. The thermal mass of masonry can be used for natural solar heating and cooling at night, resulting in fuel savings. Masonry is produced locally in most parts of the world, reducing energy lost in shipping. Brick's durable, weather-resistant character gives it tremendous longevity, and very little waste is produced on the construction site. When a building is demolished, the bricks can be cleaned and recycled for other purposes, such as landfill on a construction site. However, producing brick masonry is energy intensive due to the necessary firing. Brick kilns use oil, gas, or coal, and improperly regulated kilns produce air pollution.

BRICK

A brick is a rectangular masonry unit that is molded from clays and shales, dried, and then fired in a kiln. The physical and chemical characteristics of the ingredients vary considerably, and this, along with the temperatures at which they are fired, accounts for the variations in color and hardness. The differences among finished bricks are also dependent on the method of molding used.

Molding Methods

There are three main methods of molding brick. The *soft mud process* uses molds into which moist clay is pressed by hand or machine into rectangular molds.

The widely used *stiff mud process* forces the mixture through a die, extruding a clay ribbon that is cut into bricks by tightly stretched wires (wire-cut brick).

The *dry-press process* uses a relatively dry mixture that is pressed into gang molds by plungers under high pressure. This process produces the most accurately formed brick.

Bricks are produced in solid, cored, and hollow units, although cored brick is classified as solid if at least 75 percent of its total cross-sectional area is solid. In hollow brick, at least 60 percent of the total cross-sectional area is solid. Most natural brick colors are in the red or buff ranges, while the commonly available surface textures are smooth, water-and sand-struck, scored, wire-cut, combed, and roughened. Brick can also be finished with a fire-bonded ceramic glaze in either a satin or gloss finish.

Brick Types

Building brick, sometimes referred to as *common brick*, is the type most widely used in construction. It is made from clay fired at about 1,850°F, usually red in color, and available in various sizes. Although there is no U.S. standard brick size, the closest thing to it is the standard building brick, which has the approximate dimensions shown in Figure 2.1.

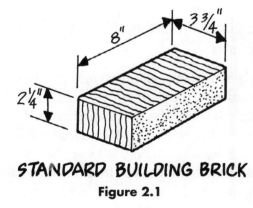

STANDARD BUILDING BRICK
Figure 2.1

Brick is specified in accordance with the use and exposure to which it will be subjected, as follows:

Grade SW (Severe weathering) Used in areas of heavy rain, snow, or continual freezing.

Grade MW (Moderate weathering) Used in areas of average rain and moderate freezing.

Grade NW (No weathering) Used in areas of minimal rain and no freezing, as in sheltered or indoor locations.

Face brick is brick that will be exposed to view, and it is made from controlled mixtures of clay or shale and carefully manufactured to produce high quality units in specific sizes, textures, and colors. Face brick, too, is specified according to exposure, and is available in SW and MW grades. In addition, it is classified according to factors affecting its appearance, as follows:

Grade FBX High degree of mechanical perfection, narrow color range, and minimum size variation.

Grade FBS Greater size variation and wide color range.

Grade FBA Nonuniform in size, color, and texture.

Hollow brick is available in SW and MW grades, and is classified by factors affecting its appearance, similar to face brick, as follows:

Grade HBX High degree of mechanical perfection, minimum size variation, and narrow color range.

Grade HBS Greater size variation and wide color range.

Grade HBA Nonuniform in size, color, and texture.

Other types of brick include the following:

Backup brick. Inferior brick used behind face brick.

Paving brick. Very hard and dense brick used in pavements.

Fire brick. Brick made with great resistance to high temperatures, as in a fireplace.

Sewer brick. Low-absorption brick for use in sewerage and storm drains.

Adobe brick. Brick made from a mixture of natural clay and straw, placed in molds, and dried in the sun. Requires protection from rain and subsurface moisture.

Nail-on brick. Flat brick generally used on interiors where solid masonry cannot be structurally supported.

Hollow brick. Brick whose net cross-sectional area is at least 60 percent of its gross cross-sectional area.

Most bricks are manufactured in standard sizes, but there are dimensional variations, based on the region and the manufacturer. Some variations are caused by the differences in clays used and the amount of shrinkage that occurs when bricks are fired.

Modular bricks have dimensions such that one or more brick courses plus the mortar joints produce courses with an exact dimension, which is usually a multiple of 4 inches. For example, an Economy-8 modular brick $3\frac{1}{2}$ inches $\times$ $7\frac{1}{2}$ inches $\times$ $3\frac{1}{2}$ inches in size laid with $\frac{1}{2}$ inch joints will produce regular courses exactly 4 inches in height and 8 inches in length. Other modular bricks include Roman ($1\frac{1}{2}$ inches high) and Economy-12 ($3\frac{1}{2}$ inches high).

END Day 1

Brick Nomenclature

Special terms are used to identify, classify, and describe brick shapes, surfaces, and placement within a wall. For example, the six surfaces of a brick are called the *face, side, cull, end*, and *beds*, as shown in Figure 2.2.

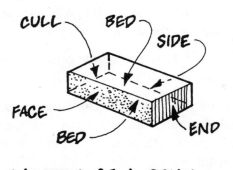

SURFACES OF A BRICK
Figure 2.2

Frequently, the mason must cut bricks into various shapes in order to fill in the spaces at corners and other places where a full brick will not fit. The more common cut shapes are shown in Figure 2.3.

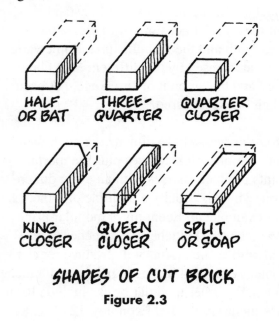

SHAPES OF CUT BRICK
Figure 2.3

Bricks may also be identified by their placement within a wall. For example, if they are laid with the end or cull (short sides) exposed, they are called *headers*; but if they are laid with the face (long side) exposed they are called *stretchers*. Various other positions are illustrated in Figure 2.4.

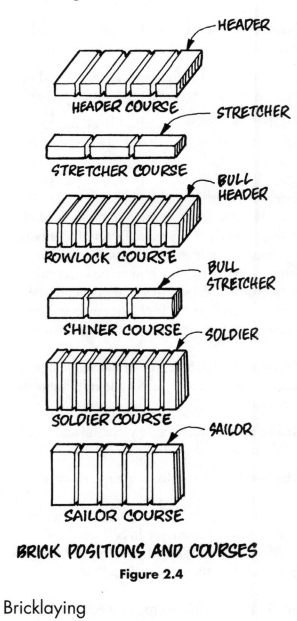

BRICK POSITIONS AND COURSES
Figure 2.4

Bricklaying

Bricklaying has always been a manual, rather than a mechanical, process, and therefore, the quality of brickwork depends to a large extent on workmanship. In this regard, there are several standard practices with which one should be familiar. For example, bricks should be laid when the temperature is between 40 and 90°F for best results.

In addition, bricks should be wetted prior to setting in order to minimize absorption of water from the mortar. Loss of water will cause the mortar to set too soon, and the bond between brick and mortar will be weak. Wetting bricks will also wash dust from the brick surface, resulting in better bond. Tests have indicated that the absorption rate of wetted brick should not exceed 0.7 ounces of water per minute. If the absorption rate exceeds this amount, the bricks should be rewetted until the desired value is achieved.

Bricks should always be set in a full bed of mortar with mortar solidly filling all vertical head joints. Standard joint thicknesses vary, although they usually range between ¼ and ½ inch. In general, three bricks plus three joints equals eight inches in height.

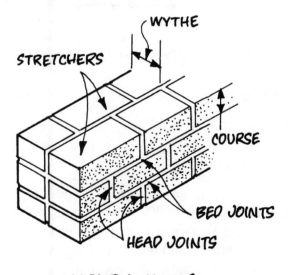

MASONRY JOINTS
Figure 2.5

Few brick walls today are laid solid. Most walls are cavity walls, in which two tiers, or wythes, of masonry are separated by two or three inches of air space. Rigid foam insulation, applied to the inside wythe, occupies about half this space, thus increasing the insulation value of the wall.

Reinforced brick masonry consists of two wythes of brick separated by a two- to four-inch space in which horizontal and vertical reinforcing bars are placed. The space is filled solidly with grout, which is a mixture of portland cement, sand, water, and sometimes pea gravel. A small amount of lime is also permitted in the grout mix. Reinforced brick masonry is much stronger than unreinforced brick masonry, for both vertical loads and lateral loads from wind or earthquake.

Brick Bonding

Brick bonding refers to the pattern in which bricks are laid to tie the wythes together into a structural unit. Some of the common bond patterns are shown in Figure 2.6. *Common bond* has a header course every sixth course, *English bond* alternates header and stretcher courses, *Flemish bond* alternates headers and stretchers in each course, and *running bond* and *stacked bond* have mesh reinforcing every sixth course. Reinforced brick masonry is generally laid in running bond or stacked bond, and as the cavity between the brick wythes contains reinforcing bars, mesh reinforcing is unnecessary.

Veneering

Veneering is the term used for exposed masonry that is attached, but not structurally bonded, to the backing. The veneer units are held in place by metal wires, clips, or anchors.

Efflorescence

Efflorescence is a white, powdery deposit on the masonry surface caused by soluble salts in the units or in the mortar. These salts are leached out by water that penetrates the masonry and results in unsightly patches of discoloration.

Efflorescence can be prevented, or at least minimized, by selecting materials free of harmful salts and by preventing water from penetrating

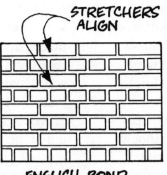

STRETCHERS ALIGN

STRETCHERS DO NOT ALIGN

FLEMISH BOND

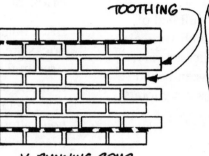

ENGLISH BOND

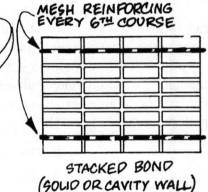

CROSS BOND
(ENGLISH, FLEMISH OR DUTCH)

HEADER EVERY 6TH COURSE

COMMON BOND
(RUNNING BOND IF NO HEADERS)

TOOTHING

1/3 RUNNING BOND
(SOLID OR CAVITY WALLS)

MESH REINFORCING EVERY 6TH COURSE

STACKED BOND
(SOLID OR CAVITY WALL)

BRICK BONDING PATTERNS

Figure 2.6

the masonry. This may be accomplished by the use of solid and tight mortar joints, capped walls, effective flashing, and adequate weather protection of the masonry during the course of construction.

When efflorescence appears, it can be removed by washing with high pressure water, by light sandblasting, or most commonly by washing with a 5 percent solution of muriatic acid in water.

Expansion Joints

Expansion joints are required in all masonry structures more than 200 feet in length, or where there are two or more wings in a building. Temperature and moisture changes cause expansion and contraction, but in masonry walls much of the movement is accommodated by the wall itself because of the flexibility in the many mortar joints. Nevertheless, to avoid harmful cracking, expansion joints such as those shown in Figure 2.7 should be used. In these joints, the sealant adheres to the two masonry surfaces to prevent air and water infiltration, while permitting movement parallel to the wall face. The compressible filler is used to maintain the required depth of the sealant.

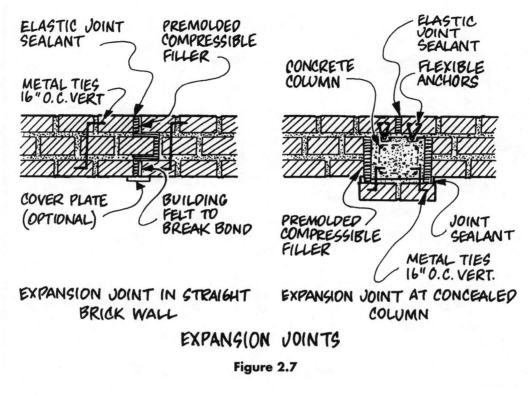

ELASTIC JOINT SEALANT

PREMOLDED COMPRESSIBLE FILLER

METAL TIES 16" O.C. VERT

COVER PLATE (OPTIONAL)

BUILDING FELT TO BREAK BOND

ELASTIC JOINT SEALANT

CONCRETE COLUMN

FLEXIBLE ANCHORS

PREMOLDED COMPRESSIBLE FILLER

JOINT SEALANT

METAL TIES 16" O.C. VERT.

EXPANSION JOINT IN STRAIGHT BRICK WALL

EXPANSION JOINT AT CONCEALED COLUMN

EXPANSION JOINTS

Figure 2.7

CONCRETE MASONRY

A wide variety of concrete masonry units are made from concrete, using either normal or lightweight aggregates. The units are manufactured by consolidating a stiff concrete mixture in steel molds. The molds are immediately removed and the units cured at an accelerated rate and dried. Units include concrete bricks, concrete blocks, concrete tile, and cast (concrete) stone. The most widely used of these masonry products is *concrete block*, which has become increasingly important as a construction material.

Concrete blocks are easy and relatively inexpensive to manufacture in a variety of patterns, textures, and colors. They are light, strong, and have good fire resistance, and they can be used for foundations, partitions, and load-bearing or non-load-bearing walls. Concrete blocks are modular, so that a nominal 8 × 8 × 16 block actually measures 7⅝ × 7⅝ × 15⅝ inches to allow for ⅜ inch mortar joints, both horizontally and vertically.

Individual blocks are manufactured with two or three cores (cells), as well as solid. As compared to solid units, two-cell concrete blocks reduce heat conduction, are lighter, and have more space for mechanical pipes and conduits.

Load-bearing concrete block is classified in two grades: N for more severe exposures and S for block requiring protection from the weather.

Concrete block walls are normally constructed one block thick and laid with staggered vertical joints (running bond). Steel reinforcing bars are often placed vertically in the middle of grout-filled cells. Horizontal reinforcement consists of either thin welded grids laid in bed joints or reinforcing bars placed in blocks with channeled webs, called bond beam blocks. Lintels over door and window openings can be made of steel or reinforced concrete, or solid-grouted lintel or bond beam blocks with horizontal reinforcing bars.

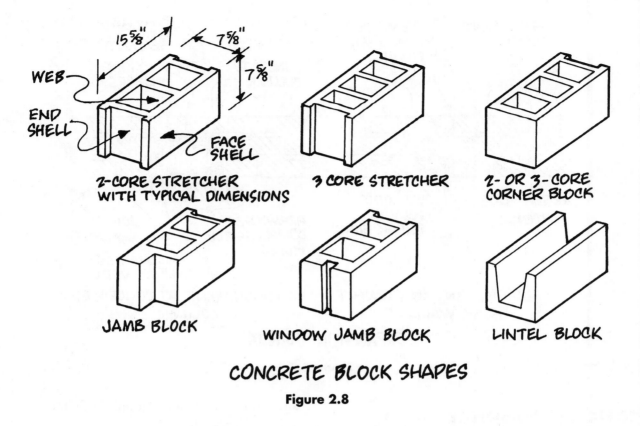

CONCRETE BLOCK SHAPES

Figure 2.8

Block walls may be left unfinished, thus displaying the color of the cement used in the manufacturing process, or they may receive a coating of cement plaster, cement paint, or clear waterproof sealer. Concrete block can also be used as a backup wall behind a facing of brick, stone, or tile.

Concrete block masonry is usually less expensive per square foot of wall than brick or stone masonry. Although concrete blocks can be saw-cut on the job, it is more economical if the architect lays out the dimensions of the building to correspond with the block module, which is generally eight inches.

END DAY 2.

STRUCTURAL CLAY TILE

Structural clay tiles are hollow, burned-clay masonry units with parallel cells. They are made from the same clays as brick, and may be load-bearing or non-load-bearing (partition tile). Clay tiles are divided into two broad groups according to function: backup tile and facing tile. They are available in a variety of textures, in natural colors (usually in the range of brick reds), or in various glazed finishes.

Structural clay tile is used for interior partitions, or, in combination with other masonry, as back-up for exterior walls. Depending on the orientation of the cells, the tile is referred to as a side-construction tile (cells horizontal) or an end-construction tile (cells vertical).

Architectural terra cotta is clay tile that is available in various colors, textures, and shapes. It is used primarily for multicolored decorative designs.

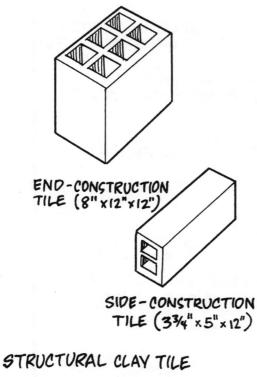

END-CONSTRUCTION
TILE (8"x12"x12")

SIDE-CONSTRUCTION
TILE (3¾"x5"x12")

STRUCTURAL CLAY TILE
Figure 2.9

Ceramic veneer is terra cotta available in large face dimensions, thin sections, and a variety of natural and glazed finishes. It is applied with a mortar setting bed, or by using metal anchors and mortar.

GYPSUM BLOCK

Gypsum blocks, often referred to as gypsum tiles, are solid or cored units manufactured from gypsum plaster. They are available in thicknesses from 2 to 6 inches, and in standard panels 12 × 30 inches in size.

Gypsum block is used for interior non-load-bearing partitions and, because of its chemical properties, for lightweight fireproofing protection. For example, two inches of gypsum block has the same fire rating as four inches of concrete block. Gypsum block cannot be used on the exterior or in areas subject to continuous

dampness. It is always set with gypsum mortar on top of a base course of water-resistant material.

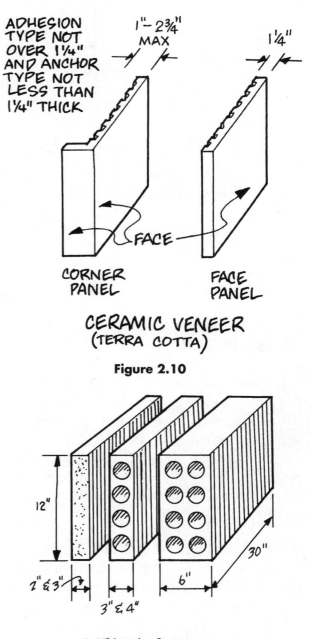

ADHESION
TYPE NOT
OVER 1¼"
AND ANCHOR
TYPE NOT
LESS THAN
1¼" THICK

1"- 2¾" MAX

1¼"

FACE

CORNER
PANEL

FACE
PANEL

CERAMIC VENEER
(TERRA COTTA)
Figure 2.10

12"

2" & 3"

3" & 4"

6"

30"

GYPSUM BLOCKS
Figure 2.11

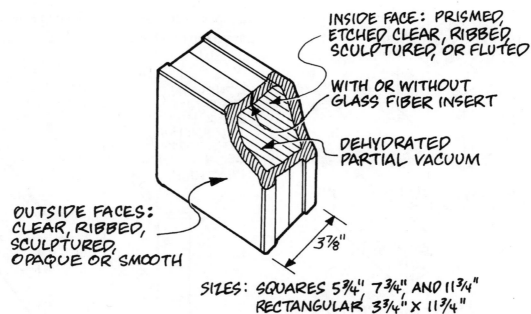

INSIDE FACE: PRISMED, ETCHED CLEAR, RIBBED SCULPTURED, OR FLUTED

WITH OR WITHOUT GLASS FIBER INSERT

DEHYDRATED PARTIAL VACUUM

OUTSIDE FACES: CLEAR, RIBBED, SCULPTURED, OPAQUE OR SMOOTH

3⅞"

SIZES: SQUARES 5¾", 7¾", AND 11¾"
RECTANGULAR 3¾" X 11¾"

GLASS BLOCK

Figure 2.12

GLASS BLOCK

Glass blocks are either solid or hollow units that have the same general chemical composition as window glass. They are used where light transmission, glare, or solar heat must be controlled, or where a specific decorative effect is desired. Glass blocks are available in several standard sizes, all of which are based on a four-inch module.

Walls of glass block are limited in length, height, and area, and they may never be used to support structural loads. The blocks are always set in a stacked bond pattern. When using glass block, one must give special consideration to the mortar mix because of the poor bond between glass and mortar. In addition, control joints are critical because of the high coefficient of thermal expansion for glass. Also critical are the joints with other materials, which should be filled with resilient expansion joint material.

Glass blocks were very popular following their introduction in the 1930s. However, due to overuse and frequent misuse, their popularity waned in the post-war period. In recent years, there has been increased interest in glass block for decorative purposes and because of the demand for greater insulation values in exterior wall materials.

STONE

Stone is one of the original natural materials used in construction by prehistoric people. As such, it has served literally as the foundation of all architectural efforts, from prehistory to the turn of the last century. Today stone is used for aggregates or as a surface finish material, in the form of veneer, paving, shingles, countertops, and various decorative items.

Classification

Stone consists of small or quarried pieces of rock, of which there are three types: *igneous*, *sedimentary*, and *metamorphic*, based on how the rock was formed. The principal building stones in each type that are commonly used in construction are listed in Table 2.1.

Type of Rock	Name of Stone
Igneous	Granite
Sedimentary	Limestone
	Sandstone
	Bluestone
	Brownstone
Metamorphic	Marble
	Soapstone
	Slate

Table 2.1

Stone Forms

Stone is available in a number of forms and used in construction in a variety of ways. Some of the most common forms are as follows:

Rough stone (fieldstone)—natural stone used decoratively

Rubble stone—irregular stone with at least one good face used for ashlar veneers, copings, sills, curbs, etc.

Dimension stone—cut stone, used for surface veneers, toilet partitions, flooring, stair treads, etc.

Flagstone—thin slabs used for paving, treads, counter tops, etc.

Monumental stone—used for sculpture, monuments, gravestones, etc.

Crushed stone—used as aggregate for concrete, asphaltic concrete, terrazzo, built-up roof surfacing, etc., as well as granular fill

Stone dust—used as filler in asphalt flooring, shingles, paints, etc.

The properties of natural stone vary considerably, and therefore, one must choose carefully with regard to strength, porosity, absorption, and permeability. It is also important to select a surface finish that will be appropriate to the stone type and suitable for the function it must perform. Surface finishes range from very rough (quarry face, split face, or sawed finish) to quite smooth (rubbed finish, honed finish, or polished finish). All stone is affected by rapid temperature change, and therefore should not be used where fire resistance is important.

Stone Masonry

Stone masonry is classified in two principal groups: *rubble masonry*, in which the stones are left in their natural rough state, and *ashlar masonry*, in which the stones are shaped and smoothed (dressed) into rectangular blocks. Stone masonry is further categorized as *coarsed*, which has continuous horizontal joints, and *uncoarsed* or *random*, which does not. A *bond stone* is a stone with its longest dimension perpendicular to the wall face to tie the wall to its backing. Several stone masonry patterns are shown in Figure 2.13.

Stone masonry may be laid in mortar, similar to brick and concrete block, or it may be in the form of thin veneer mechanically anchored to the backup wall or frame.

Stone masonry should be set with nonstaining portland cement mortar, and care should be taken to avoid moisture penetration of the stone. This is usually accomplished by damp-proofing the backup material. There are many varieties of anchors and ties used to secure stone veneer or stone trim, and several of these are shown in Figure 2.14.

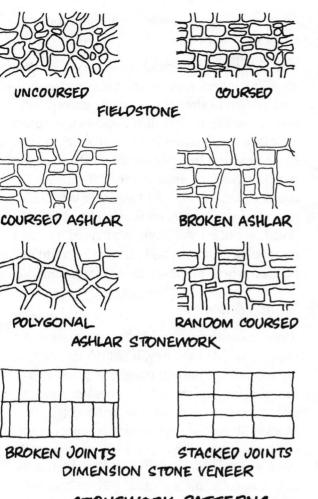

UNCOURSED   COURSED
FIELDSTONE

COURSED ASHLAR   BROKEN ASHLAR

POLYGONAL   RANDOM COURSED
ASHLAR STONEWORK

BROKEN JOINTS   STACKED JOINTS
DIMENSION STONE VENEER

STONEWORK PATTERNS
Figure 2.13

END DAY 3

MORTAR

All masonry units are bonded with mortar, the purpose of which is to join the units to each other, or to their supporting members, while preventing moisture penetration of the joints. Mortar is composed of varying quantities of portland cement, sand, lime, and water. Mortar sand should be clean and free from organic material, and its grading is based on the thickness of the mortar joint. Lime putty or hydrated lime is used to improve workability and water retentivity, although it reduces the strength of the mortar. Admixtures are permitted, but except for powdered coloring agents, they are not recommended where high bond strength is desired.

Masonry cement or mortar cement may be used instead of, or in addition to, portland cement, in which case no lime may be added. However, only portland cement-lime mortar should be used where high strength and low permeability are required.

Mortars may be tested for slump or flow, similar to concrete, in order to insure a proper mix. Mixes that have lost water through evaporation

MORTAR PROPORTIONS FOR UNIT MASONRY

MORTAR	TYPE	PROPORTIONS BY VOLUME (CEMENTITIOUS MATERIALS)								AGGREGATE MEASURED IN A DAMP, LOOSE CONDITION
		Portland Cement or Blended Cement	Masonry Cement[1]			Mortar Cement[2]			Hydrated Lime or Lime Putty	
			M	S	N	M	S	N		
Cement-lime	M	1	—	—	—	—	—	—	$\frac{1}{4}$	
	S	1	—	—	—	—	—	—	over $\frac{1}{4}$ to $\frac{1}{2}$	
	N	1	—	—	—	—	—	—	over $\frac{1}{2}$ to $1\frac{1}{4}$	
	O	1	—	—	—	—	—	—	over $1\frac{1}{4}$ to $2\frac{1}{2}$	
Mortar cement	M	1	—	—	—	—	—	1	—	Not less than $2\frac{1}{4}$ and not more than 3 times the sum of the separate volumes of cementitious materials.
	M	—	—	—	—	1	—	—	—	
	S	$\frac{1}{2}$	—	—	—	—	—	1	—	
	S	—	—	—	—	—	1	—	—	
	N	—	—	—	—	—	—	1	—	
Masonry cement	M	1	—	—	1	—	—	—	—	
	M	—	1	—	—	—	—	—	—	
	S	$\frac{1}{2}$	—	—	1	—	—	—	—	
	S	—	—	1	—	—	—	—	—	
	N	—	—	—	1	—	—	—	—	
	O	—	—	—	1	—	—	—	—	

[1]Masonry cement conforming to the requirements of U.B.C. Standard 21-11.
[2]Mortar cement conforming to the requirements of U.B.C. Standard 21-14.

Table 2.2

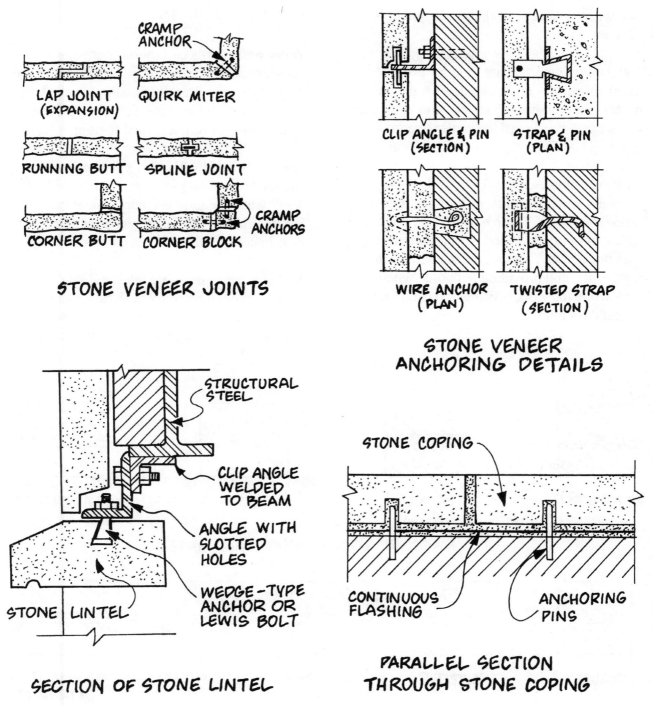

CRAMP ANCHOR

LAP JOINT (EXPANSION)

QUIRK MITER

RUNNING BUTT

SPLINE JOINT

CORNER BUTT

CORNER BLOCK

CRAMP ANCHORS

STONE VENEER JOINTS

CLIP ANGLE & PIN (SECTION)

STRAP & PIN (PLAN)

WIRE ANCHOR (PLAN)

TWISTED STRAP (SECTION)

STONE VENEER ANCHORING DETAILS

STRUCTURAL STEEL

CLIP ANGLE WELDED TO BEAM

ANGLE WITH SLOTTED HOLES

WEDGE-TYPE ANCHOR OR LEWIS BOLT

STONE LINTEL

SECTION OF STONE LINTEL

STONE COPING

CONTINUOUS FLASHING

ANCHORING PINS

PARALLEL SECTION THROUGH STONE COPING

Figure 2.14

may be retempered (water added); however, mortar should not be used more than three hours after it is mixed.

A table of mortar proportions reproduced from the Uniform Building Code is shown in Table 2.2. For masonry that is load-bearing and/or exposed to the weather, type M or S mortar is generally specified. Types N and O mortars are used where a lesser compressive strength is required.

High-strength mortars are used for preassembled masonry units, and as such they have greater compressive and tensile strength. These mortars consist of portland cement, sand, a high-bond additive, a plasticizer, and water.

In severe climates, special precautions may be necessary when masonry work is performed during the winter. These may include protecting the masonry units and mortar ingredients from the weather, heating the water used in the mortar mix, and providing temporary enclosures while the work is being done and for several days after.

MORTAR JOINTS

The exterior surface of mortar joints, which is exposed to the weather, is finished to make the masonry more waterproof and/or to achieve a specific aesthetic appearance. Joints may be made with a trowel, a special jointing tool, or a raking device. In each case, the joint should be entirely filled with mortar to begin with, and the face of the joint should be smooth and dense when the joint is completed. Shown in Figure 2.15 are several common masonry joints used in construction.

Masonry accessories are readily available for each type of masonry unit; they include anchors, ties, reinforcing, fillers, wire mesh, and so on, as well as various accessories for

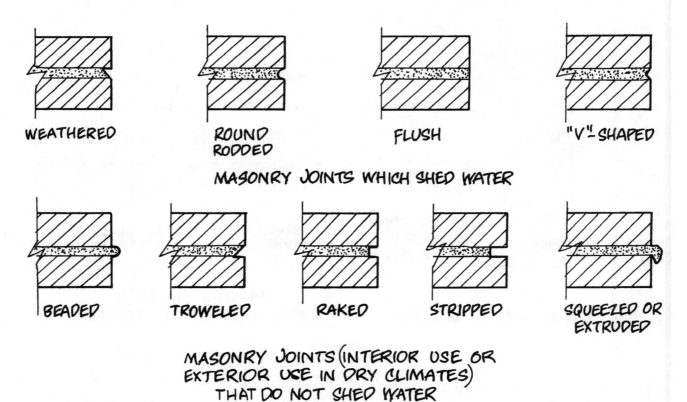

WEATHERED   ROUND RODDED   FLUSH   "V"-SHAPED

MASONRY JOINTS WHICH SHED WATER

BEADED   TROWELED   RAKED   STRIPPED   SQUEEZED OR EXTRUDED

MASONRY JOINTS (INTERIOR USE OR EXTERIOR USE IN DRY CLIMATES) THAT DO NOT SHED WATER

COMMON MORTAR JOINTS

Figure 2.15

installing masonry ceilings, as shown in Figure 2.16.

CONCLUSION

The use of masonry has been called an anachronism in the modern world of construction. As most masonry work is performed very much as it has been for centuries, there is no doubt that it is labor intensive and expensive. However, recent innovations, including prefabricated units, preassembled panels, and high bond adhesives, have led to decreased costs and expanded use.

There is also a psychological factor associated with masonry work; its warmth, scale, and handcrafted appearance are pleasing and cannot be replaced by factory-produced materials. Perhaps one day the stone wall and brick-paved patio will go the way of the cobbled street, but that day still appears to be in the distant future.

END DAY 4.

STRAP ANCHORS

DOVETAIL ANCHORS

CRAMP ANCHORS PIN THREADED DOWEL

HANGERS

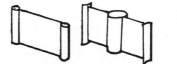

EXPANSION JOINTS
WATER STOPS

MASONRY ACCESSORIES
Figure 2.16

1. In a situation that requires a lightweight, fire-resistant interior partition, which of the following materials would be preferred?

   A. Concrete block

   B. Glass block

   C. Gypsum block

   D. Common brick

2. If a project required a nonbearing exterior wall to minimize solar heat gain while transmitting some light, which of the following materials would be appropriate?

   A. Glass block

   B. Face brick

   C. Modular brick

   D. Grade MW brick

3. Face brick differs from building brick in that it is generally

   A. more resistant to severe weathering.

   B. harder and more durable.

   C. more uniform in size and color.

   D. available in a variety of sizes.

4. To construct an eight-inch-thick masonry wall, concrete block might be selected instead of brick because it is

   I.   less expensive.

   II.   easier to handle.

   III. easier to reinforce.

   IV.   faster to erect.

   V.   more attractive.

   A. I and IV   C. I, II, and IV

   B. I and III   D. II, III, and V

5. Which of the following has little effect in preventing efflorescence?

   A. Using solid, tight mortar joints

   B. Wetting the masonry units prior to setting

   C. Furnishing adequate weather protection during construction

   D. Providing a continuous weatherproof cap on walls

6. A mortar mixture of portland cement, sand, and water, but no hydrated lime, would probably produce a mortar with

   A. greater strength and workability.

   B. insufficient bonding strength.

   C. little resistance to moisture penetration.

   D. insufficient water retention.

7. The considerable variation found in natural brick colors is caused by the

   I.   preference of the manufacturer.

   II.   clays from which they are made.

   III. pigmented admixtures used.

   IV.   temperatures at which they are fired.

   V.   length of time they are stored.

   A. II only   C. I, III, and IV

   B. II and IV   D. II, III, and V

8. A masonry course in which the long dimension of the exposed ends is placed vertically is called a

   A. header.   C. stretcher.

   B. soldier.   D. rowlock.

9. During a weekly job-site visit, the architect notices that a mason has used troweled joints on an exposed exterior masonry wall, instead of the weather-struck joints that were specified. In the interest of practicality, the architect should

    A. accept the change, because there is little difference in the two joints.

    B. accept the change, because there is no practical way to modify troweled joints.

    C. reject the work and have the mason trowel additional mortar over every horizontal joint.

    D. reject the wall and have the wall rebuilt.

10. The most reasonable alternative solution to the problem stated in the previous question would be to

    A. paint the entire wall with a waterproof sealer.

    B. spray the entire wall with a clear lacquer.

    C. fill each joint with expansion joint filler.

    D. plaster the masonry wall.

WOOD

INTRODUCTION

Wood is unique, as it is the only building material that grows and is renewable. Each year a new layer of wood develops naturally on a tree, with no need for human intervention. Therefore, with intelligent planning and forestry management, wood should remain available to us forever.

Because wood is organic, it does not have the precise properties of a manufactured product. On the contrary, wood has numerous and varied imperfections, such as checks, knots, pitch pockets, and shakes. To overcome some of these natural defects, wood used in construction is often modified to create more precisely engineered products, such as glued laminated lumber, panel products, and chemically treated wood with improved resistance to fire and decay.

HISTORY

Discovery and Early Uses

Wood, mud, and stone were the first natural building materials used by prehistoric people. As humans settled into farming communities in the Neolithic era, they cut trees and used the branches and trunks for protective walls, lintels, and roofs. The pithouse, longhouse, and tipi were some of the earliest structures using saplings or logs to create shelter. Wood buildings were developed by every civilization with access to forests. Where sun-dried mud brick or stone were more readily available, masonry load-bearing walls supported timber floors and roofs.

The earliest wood joining was done by cutting or notching the wood to interlock the parts. Pins made of very hard wood were inserted into the joints. Wood pins were eventually replaced with crude nails and clamps, first of bronze, then iron. The addition of primitive forms of adhesives from asphalt, animal, and vegetable glues was limited in use. The Romans developed a method for forging nails similar to that used in the American colonies.

The Japanese are well-known for their appreciation of wood and their ability to bring out its best, natural qualities in everyday objects, furniture, and building construction. Kansai, the region around Kyoto, Japan, is renowned for its beautiful all-wood buildings, including five-story pagodas built with intricate joinery and the beam and bracket method. A beautiful example, the Hōryū-ji temple and pagoda, still stands in Nara.

In the Middle Ages, half-timbering became a standard method of construction in northern Europe, where forests were plentiful. Timber frames were joined together with pegged mortise and tenon joints, scarf joints, lap joints, and diagonal bracing to keep the building from racking. The timber frames formed the structure of the buildings, whereas the spaces between the timbers were filled either with wattle and daub, brick, or rubble and faced with plaster on the interior and exterior of the building. Wood wainscots were often added to the interior as a layer of insulation. Public buildings often displayed great imagination and beauty in their design, such as the Norwegian stave churches of the 12th century and Westminster Hall in London—one of the finest examples of refined timber joinery.

Europeans brought their expertise in heavy timber frame construction and wood joinery to colonial America. Nails were rare and expensive, and were therefore used only for door and window construction and for attaching siding. Nails were forged until around 1720, when an easier method for making nails from wire was invented.

Major Developments

A great change in wood construction occurred with the development of machinery. The invention of the water-powered saw mill in the beginning of the 19th century made possible the production of dimension lumber in large quantities for commercial purposes. Industrial mills transformed felled trees into boards and squared timbers in a fraction of the time it took men to hand-hew lumber with a two-person pit saw. Rotary cutting made the production of wood veneer easier and, in combination with the development of improved natural and synthetic adhesives, led to the invention of plywood.

In 1833, Augustine Taylor developed a new, lighter method of wood construction while building St. Mary's Church in Chicago. Taylor replaced heavy timbers with a series of closely placed thin wood studs that ran from the foundation up to the eaves of the building, and attached the studs to wood beams with cheap, machine-made nails. The intermediate floor structure was simply nailed to the studs. Skilled carpenters nearby ridiculed his experiment, saying that it looked no more substantial than a balloon and would surely blow over in the wind. However, balloon framing, which never lost its nickname, proved to be an economic breakthrough for a rapidly growing American population bent on expanding westward.

Lightweight wood frame construction was made more efficient and fireproof when the platform frame was developed, the universal standard in construction today.

The first glue-laminated timber forms were a series of arches constructed at the beginning of the 20th century in Europe. Experiments with laminated wood were made for structural members in aircraft during World War I, and this technology soon made its way into the construction industry. The first laminated arches in the U.S. were made for a building erected by the Forest Products Laboratory in the 1930s.

Notable Movements, Architects, and Structures

The expanding railroad network in the United States around 1900 helped to popularize several uniquely American housing styles built in half-timber and light wood frame construction, especially in southern and western states. Pre-cut architectural details for the Queen Anne Victorian, such as spindlework and geometric cut-outs, were made available throughout the nation. In the early 20th century, all-wood bungalows could be ordered from companies like Sears and Roebuck, which sold homes through its mail-order service.

The Arts and Crafts Movement, a reformist movement opposing eclectic turn of the century styles, was in part inspired by the simple hand-crafted forms seen in Japanese architecture. The Greene brothers (Charles and Henry) developed a version of the Craftsman bungalow with an exquisitely detailed exposed wood structural system reminiscent of Japanese beam and bracket construction. Their most well-known work is the Gamble House, built in 1908 in Pasadena, California.

The Shingle style, a free-form, sculptural housing style popular for a brief period of time at the end of the 1800s, inspired a modernist revival with more angular forms in the 1960s and 1970s. This minimalist style, part of the northern California vernacular, served as inspiration for the Sea Ranch Condominiums, designed by Charles Moore in 1965.

Recent Trends

Wood is experiencing renewed popularity as a building material in light of its naturally sustainable qualities, which can be enhanced by sensitive design and construction practices. In 1980, E. Fay Jones built a remarkable lightweight wood structure, the Thorncrown Chapel, in a forest near Eureka Springs, Arkansas. He framed large expanses of glass in a forest of slim two-inch lumber. In order to protect the virgin site, the materials for construction were transported to the site by hand.

Part of wood's appeal is that it is the only renewable structural material. Architects can support sustainable forestry practices by specifying products from local, certified forests and reducing the amount of steel used as fasteners. Construction site waste can be reduced by designing buildings that use full-size sheets of plywood and standard lengths of lumber. There is a growing industry in the recycling of old building materials.

Wood is also transforming as a building material. It can be impregnated or laminated with new forms of rubber, plastics, and adhesives. It can be permanently bonded to sheet metal, and can be shaped into forms needed to meet structural requirements.

TERMINOLOGY AND CLASSIFICATION

The terms *wood, lumber,* and *timber* are often used interchangeably, but each term has a distinct meaning.

1. *Wood* is the hard fibrous substance lying beneath the bark of trees.

2. *Lumber* is wood that has been sawn into construction members.

3. *Timber* is lumber that is five inches or larger in its least dimension.

Wood is classified as *softwood* or *hardwood*, depending on the type of tree from which it comes. Softwoods, such as pine, fir, and spruce, come from needle-leaved conifers, which are evergreen. Hardwoods, such as maple, oak, and sycamore, come from broad-leaved deciduous trees, which shed their leaves annually.

These terms are botanical, not structural, and they do not necessarily indicate the relative hardness or strength of a particular wood. For example, Douglas fir, a softwood, is much harder and stronger than basswood, which is a hardwood. And balsa, one of the most spongy and lightweight woods, is also a hardwood. Softwoods are used structurally in general construction for framing, sheathing, bracing, and so on, while hardwoods are mainly used for flooring, paneling, interior trim, and furniture.

CHARACTERISTICS

Wood is the handiest and most accommodating of all construction materials. It is available almost everywhere, readily transported, and easily worked with simple carpenters' tools. Wood-framed structures are strong, durable, and usually lower in cost than comparable concrete, masonry, or steel structures.

Although wood is combustible, heavy timber construction can resist fire better than unprotected steel construction. In addition, wood construction has great popular appeal, as light wood framing requires relatively little technical knowledge, and it may easily be worked with small, simple tools. Wood also has a unique beauty and warmth that people have enjoyed through the ages.

Wood consists of approximately 70 percent cellulose and 18 to 28 percent lignin, which is the adhesive imparting strength to the wood. The remainder is made up of minerals and extractives, which give wood its color, odor, and resistance to decay.

END DAY 1

STRENGTH OF WOOD

Wood is generally stronger in compression than tension. Unlike most other building materials, the strength of wood is not the same in every direction: for both tension and compression, wood is much stronger when the load is applied parallel to the grain than perpendicular to the grain. In fact, the strength of wood in tension perpendicular to the grain is so low that this type of loading can easily cause the wood to split and should therefore be avoided.

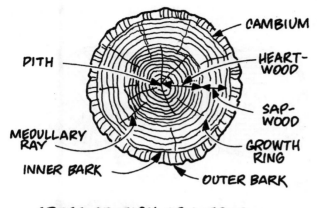

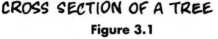

CROSS SECTION OF A TREE
Figure 3.1

For shear, wood is very strong perpendicular to the grain and relatively weak parallel to the grain. Therefore, horizontal shear stress (parallel to the grain) is often a design consideration, while vertical shear stress (perpendicular to the grain) is not.

SEASONING OF WOOD

Making wood suitable for construction involves more than simply cutting down a tree, shaking out the squirrels, and sawing the wood to size. Wood in the living tree is *green*; that is, it contains a large amount of water. If green lumber is used in construction, it will shrink as it dries out, which can cause a variety of problems. To minimize shrinkage, lumber should be *seasoned* (dried) before installation, until it reaches the moisture content it will have in service. This can be accomplished either by *air drying*, which takes several months and leaves 10 to 20 percent moisture in the lumber, or *kiln drying*, which takes only a few days and leaves less than 10 percent moisture. Framing lumber is considered seasoned if its moisture content is 19 percent or less.

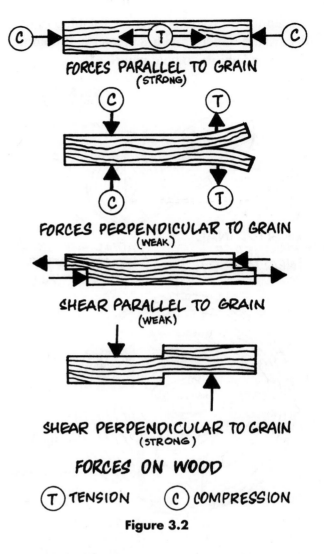

FORCES ON WOOD
Figure 3.2

Even with seasoned lumber, however, moisture changes can occur. Dry members in a humid space will absorb moisture and conversely, members with a high moisture content will lose moisture in a dry environment. Between zero moisture content and the *fiber saturation point* (about 30 percent), wood swells as it absorbs moisture and shrinks as it loses moisture. Consequently, construction details should allow for some shrinkage or swelling of lumber members.

Not only does seasoned lumber have less tendency to shrink, but it is also stronger, stiffer, and lighter in weight than green, unseasoned lumber. In addition, seasoned lumber is less susceptible to warping; more resistant to fungi, decay, and insects; and has greater nail-holding power and ability to hold paint.

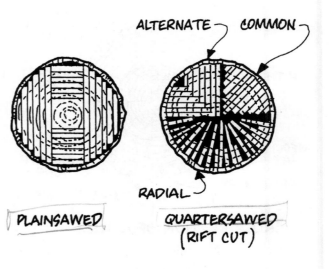

LUMBER CUTTING

Figure 3.4

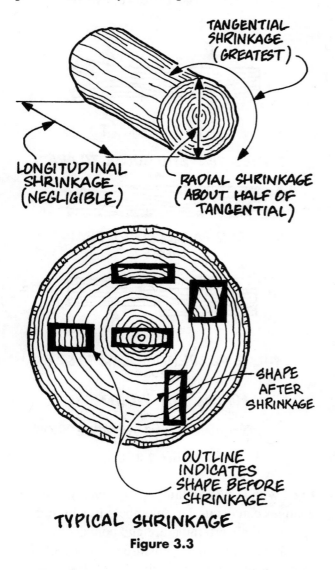

TYPICAL SHRINKAGE

Figure 3.3

CUTTING AND SAWING LUMBER

Shrinkage, distortion, and warpage of lumber depend partially on the way lumber is cut from a tree. Wood shrinks most in the direction of the annual growth rings (tangentially); less across these rings (radially); and very little parallel to the grain (longitudinally).

Lumber can be cut from a log in two different ways: tangent to the annual rings, called *plainsawed* in hardwoods and *flat-grained* or *slash-grained* in softwoods; and radially to the rings, called *quartersawed* in hardwoods and *edge-grained* or *vertical-grained* in softwoods. *Rift-grained* and *comb-grained* are other terms sometimes used for quartersawed.

In actual practice, however, this is not strictly true, and so lumber is classed as quartersawed if the grain is 45° to 90° to the wide face and plainsawed if the grain is 0° to 45° to the wide face.

The characteristics of plainsawed lumber include:

1. Distinct grain pattern
2. May twist, cup, and wear unevenly
3. Tends to have raised grain
4. Shrinks and swells more in width, less in thickness
5. Less waste in cutting, and therefore less expensive

Quartersawed lumber has the following characteristics:

1. Relatively even grain pattern
2. Wears evenly with less warpage
3. Shrinks and swells more in thickness, less in width
4. More waste in cutting and therefore more costly

WOOD DEFECTS

There are a variety of defects that affect the strength, appearance, use, and grading of lumber. Defects may be natural or caused by manufacturing. In addition to these defects, wood can be damaged by insects, decayed by fungus, and of course, destroyed by fire.

Natural Defects

Natural defects are those resulting from natural causes. Some of these are listed below:

Knot—branch embedded in a tree and cut through in manufacture

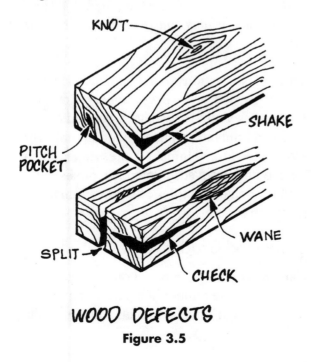

WOOD DEFECTS
Figure 3.5

Peck—pitted area sometimes found in cedar and cypress

Pitch pocket—opening between growth rings containing resin

Shake—lengthwise grain separation between or through growth rings

Manufacturing Defects

Manufacturing defects are those arising from the seasoning or processing of lumber. Some of these are listed below:

Check—lengthwise grain separation caused by seasoning

Split—lengthwise separation of wood extending from one face to another

Wane—lack of wood on the edge or corner

Warp—shrinkage distortion of a plane surface; includes bow, crook, cup, and twist

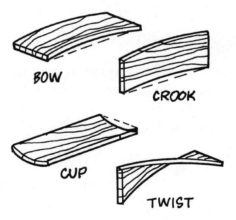

TYPES OF WARPAGE
Figure 3.6

GRADING LUMBER

As products of nature, no two pieces of wood are alike: they vary in strength, appearance, and other characteristics. In order to establish uniform standards of quality, lumber is graded for appearance or strength depending on its end use.

Grading for strength (stress grading) is done either visually or by machine, using standardized lumber grading rules. Visual grading is based on the number, size, type, and location of visible defects, such as knots, shakes, and wane. The grade assigned to a piece of lumber is then stamped on it, along with the mill identification and other information. During construction the grade is easily verified by checking the grade mark.

The best grades of lumber are practically free from imperfections, while in each lower grade, the imperfections increase in quantity.

In machine stress rating, lumber is subjected to bending, and the values of modulus of elasticity (E) and allowable fiber stress (f) are computed and automatically stamped on each piece.

Softwood Grades

Guidelines for softwood grading have been established by the National Bureau of Standards according to use, size, and manufacturer. The classification of principal uses is as follows:

Yard lumber is used in general construction.

Factory and shop lumber is used for remanufacture into products such as sash and doors, and the grades are based on the amount of clear wood in each piece.

There are three lumber size classifications: *boards, dimension lumber,* and *timbers,* as shown in Figure 3.7.

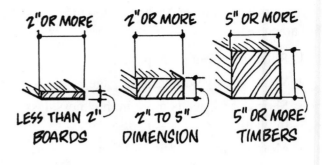

LUMBER SIZE CLASSIFICATION
Figure 3.7

Boards are graded for appearance and used as siding, subflooring, and trim. They are classified as *select* and *common*. Select lumber has a good appearance and is graded from A to D; A and B grades are of the highest quality and are suitable for natural finishes, while C and D are considered paint grades. Common lumber has more blemishes than select and is classified into five grades of descending quality, from number 1 to number 5.

Dimension lumber and timbers are called *structural lumber*, used for load-bearing members, and graded for strength. Dimension lumber is further classified according to its size as *joists and planks, light framing,* and *decking*. Timbers include *beams and stringers* and *posts and timbers*.

Three classifications based on manufacture are *rough, dressed,* and *worked lumber*. Rough lumber has visible saw marks, dressed or surfaced lumber is planed smooth to uniform sizes, and worked lumber is dressed and then tongue-and-grooved, shiplapped, or shaped to a pattern. Some examples of worked lumber are shown in Figure 3.8.

Hardwood Grades

Hardwood grades are based on the amount of clear, usable lumber in a piece. Standard lengths vary from 4 to 16 feet, and the standard

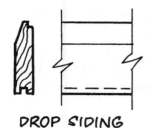

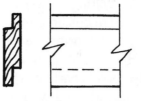
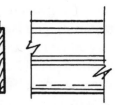

DROP SIDING   FINISH FLOORING   SHIPLAP SHEATHING FLOORING & SIDING   EXPOSED ROOF SHEATHING

WORKED LUMBER

Figure 3.8

grades are firsts, seconds, selects, sound wormy, and numbers 1, 2, 3A, and 3B common.

Lumber Sizes

All lumber is specified according to its nominal dimensions, which are the rough or unfinished sizes. The dressed, surfaced or finished size is always smaller due to the seasoning and surfacing of lumber before its use. For example, a nominal 2″ × 4″ is actually 1½ by 3½ inches in size.

Lumber is measured, computed, and priced in board feet. This is a conventional, standard unit in which a board foot is defined as a nominal 1″ × 12″ board one foot in length.

END DAY 2.

PLYWOOD

Description

Plywood is a manufactured wood panel consisting of several thin wood veneer sheets (plies) that are permanently bonded together with adhesive under high pressure, with the grain of each ply perpendicular to the grain of the adjacent plies. The center ply is known as the *core*, while the outer plies are called *face* and *back*. When there are five or more plies, the additional sheets, called *crossbands*, are located

between, and placed perpendicular to, the core and the outside plies. There is generally an odd number of plies (3, 5, or 7) that gives a balanced effect, so that the panel tends to remain flat with changes in moisture content. Some panels are made with cores of solid lumber or particle board instead of veneer.

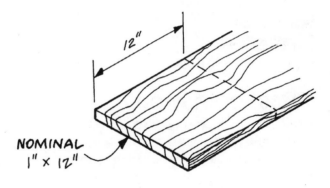

12″

NOMINAL 1″ × 12″

BOARD FOOT MEASURE
Figure 3.9

Plywood veneers are produced by rotary cutting, as in Figure 3.11. For hardwood face veneers where grain appearance is important, the veneers may be quarter sliced or plain sliced.

Plywood is produced in standard 4 by 8 foot sheets, with the grain of the face and back veneers parallel to the 8-foot dimension. Thicknesses range from 5/16 inch to 1⅛ inches.

Plywood provides a strong backing for finish materials, and is widely used in construction as wall and roof sheathing, subflooring, underlayment, and formwork. Because of its great strength and stiffness, plywood is particularly effective in resisting lateral loads from wind or earthquake, in the form of roof and floor diaphragms and shear walls.

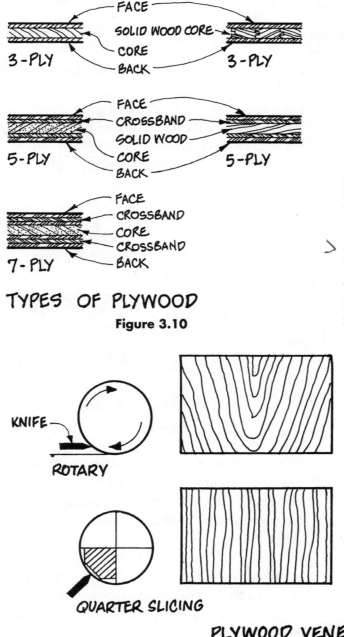

TYPES OF PLYWOOD

Figure 3.10

The advantages of plywood over sawn lumber are its great strength in both directions, greater resistance to shrinking and splitting, and less warpage. In addition, a sheet of plywood covers a large surface area with relatively few pieces of material.

Classification

Plywood is classified as *interior* or *exterior*, depending on the type of adhesive used: moisture-resistant for interior use and waterproof for exterior use. Plywood is also classified as softwood or hardwood, depending on the species of the face veneers. Plywood used in construction is primarily softwood.

Plywood is graded according to the quality of the veneers, with A being the best and D the poorest. The grade of a plywood panel is designated by two letters representing the veneer grade on the face and back; for example, an A-D panel has an A-grade face and a D-grade back. The grade is marked on each panel for easy identification. For exterior plywood, the inner plies must be grade C or better, while interior plywood may have D-grade inner plies.

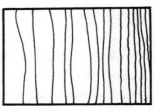

PLYWOOD VENEER CUTTING

Figure 3.11

The species of wood used for face and back veneers is identified on each panel by a group number, ranging from group 1 species (the strongest and stiffest) to group 5 species (the weakest).

Some unsanded plywood panels used for structural purposes, such as roof sheathing, subflooring, and wall sheathing, are not marked with the grade and species group. Instead, they are marked with a *span rating* consisting of two numbers; the first denotes the maximum recommended roof span in inches, and the second the maximum recommended span for subflooring. For example, a panel identified as 32/16 can be used as roof sheathing over rafters spaced 32 inches on center or as subflooring over joists spaced 16 inches on center. In addition to plywood, such panels may also be made of composites and oriented strand board.

Two typical grade marks used by mills belonging to the American Plywood Association (APA) are in Figure 3.12, by permission of the American Plywood Association.

Hardwood plywood is more expensive than softwood plywood, and is used primarily for decorative applications, such as paneling, cabinet work, face veneers on doors, etc. The grading and sizing of hardwood plywood is different from that of softwood plywood previously described.

The way in which hardwood plywood panels are matched affects the finish appearance. In *book matching*, also called *edge matching*, every other sheet is turned over. In *slip matching*, adjacent sheets are joined side by side without turning, thus repeating the grain pattern. Purposely unmatched veneers are called *random matching*.

Sanded Grades

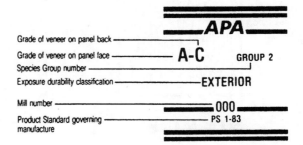

Grade of veneer on panel back
Grade of veneer on panel face — **A-C**   GROUP 2
Species Group number
Exposure durability classification — **EXTERIOR**
Mill number — **000**
Product Standard governing manufacture — PS 1-83

(Also available in Groups 1, 3 and 4)

Unsanded Grades

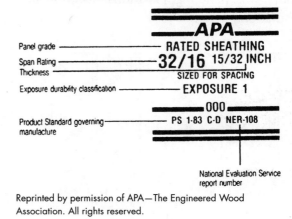

Panel grade — **RATED SHEATHING**
Span Rating — **32/16** 15/32 INCH
Thickness — SIZED FOR SPACING
Exposure durability classification — **EXPOSURE 1**
Product Standard governing manufacture — **000** — PS 1-83 C-D NER-108

National Evaluation Service report number

Figure 3.12

In addition to conventional construction plywood, other special types include overlay (covered with resin-fiber overlay); marine; prefinished (stained and ready to use); and patterned sheets (grooved, roughsawn, etc.). The major uses of these special plywoods are for furniture, cabinets, flooring, exterior siding, and interior wall finishes.

MISCELLANEOUS PANELS

Several types of pressed fiber boards are used in construction. In general, these consist of wood or other fiber mixed with a binder and

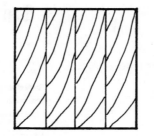

BOOK MATCH   SLIP MATCH   RANDOM MATCH

Figure 3.13

pressed into a flat sheet. They are commonly used for nonstructural or semistructural purposes, such as concrete forms, cabinets, doors, wall paneling, partitions, decking, and insulation.

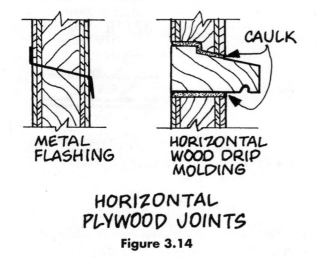

METAL FLASHING   HORIZONTAL WOOD DRIP MOLDING

CAULK

HORIZONTAL PLYWOOD JOINTS

Figure 3.14

Hardboard is made from wood fibers that have been highly compressed under heat and pressure into dense, durable boards. It is available in two categories: basic and prefinished, and in three types: tempered, standard, and service standard. These are available in 4 by 8 foot sheets that are ⅛ inch to ⅜ inch thick.

Prefinished hardboard is available in a variety of patterns, textures, and finishes, such as baked enamels, plastic laminates, etc.

Prefinished hardboard is used for exterior siding, soffits, interior walls, ceilings, cabinet work, pegboards, and acoustical treatment.

Fiberboard is manufactured from waste paper, wood pulp, and fibers. It is used for acoustical tile, sheathing, and interior wall finishes. The standard size is 4 feet by 8 feet, in thicknesses from ½ inch to 1 inch. Fiberboard roof insulation and laminated fiberboard decking are manufactured in 2-by-4-foot sheets.

Flakeboard is composed of large wood flakes bonded together with synthetic resins under pressure. Flakeboard is lightweight and has good insulation value and acoustical properties, but its surface is easily damaged. Its main use is as an insulating backup material or as an acoustical material on upper walls and ceilings.

Particleboard is dry-formed of wood particles bonded together with synthetic resin. It is used primarily as core stock for plastic laminate or hardwood veneers and used in the manufacture of furniture, cabinets, countertops, wall paneling, and doors of all types.

Beadboard is an insulating board consisting of a core of small, expanded polystyrene beads with heavy paper laminated to both sides. The major use of beadboard is as an insulating material, such as perimeter insulation on foundation walls. Its insulation value is less than

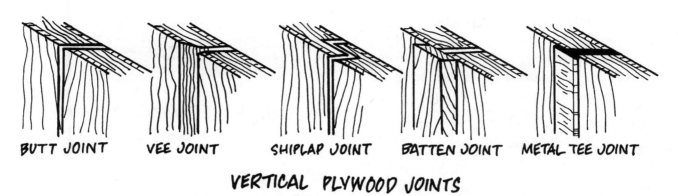

BUTT JOINT   VEE JOINT   SHIPLAP JOINT   BATTEN JOINT   METAL TEE JOINT

VERTICAL PLYWOOD JOINTS

Figure 3.15

that of extruded polystyrene insulation of the same thickness.

Formica & Wilson Art.

Plastic laminates consist of a base of phenolic-resin impregnated kraft paper over which a patterned sheet is applied. Over all this is laid a sheet of paper treated with melamine resin. These layers are then cured with intense heat and pressure to fuse the material into a thin sheet which can be applied to plywood or other backing with adhesive. Popularly known by trade names such as Formica, the material has found popular acceptance for countertops, wall coverings, and furniture.

END DAY 3

GLUED LAMINATED LUMBER

Glued laminated (*glulam*) structural members are fabricated from layers of wood that are bonded with adhesives, in which the grain of all layers is approximately parallel longitudinally. The usual thickness of laminations is 1½ inches, although ¾-inch laminations are used to achieve small radius curves. Members are produced in several standard widths and a great range of depths, up to about 75 inches.

Glued laminated members are factory produced under rigidly controlled manufacturing conditions, resulting in members that are superior to sawn lumber in several ways. For the same

size member, they can span greater distances and support heavier loads. Glulams are more weather-resistant; they are consistent in size, appearance, and strength; and they are more dimensionally stable than solid timber.

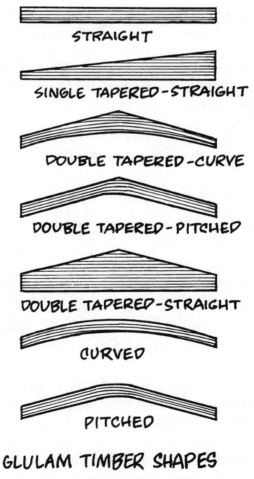

STRAIGHT

SINGLE TAPERED-STRAIGHT

DOUBLE TAPERED-CURVE

DOUBLE TAPERED-PITCHED

DOUBLE TAPERED-STRAIGHT

CURVED

PITCHED

GLULAM TIMBER SHAPES

Figure 3.16

Glulam members are available in three appearance grades:

Industrial is the grade used where appearance is not a prime concern.

Architectural is the grade used where appearance is an important consideration.

Premium is the top grade, specified where appearance is of primary importance. It is the most expensive grade and arrives at the site fully wrapped for protection.

In specifying laminated timbers, one must designate the type of adhesive (interior or exterior), the stress grade required, and the appearance grade desired.

In general, individual laminations are not the full length of the member. Therefore, they must be joined end to end in order to produce a member of the required length. The two most common types of joints are *scarf joints* and *finger joints*, as shown in Figure 3.17.

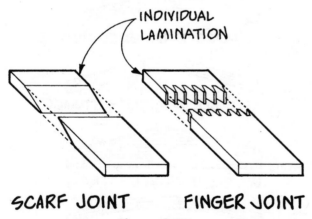

INDIVIDUAL LAMINATION

SCARF JOINT   FINGER JOINT

Figure 3.17

Laminated decking is manufactured from layers of kiln-dried lumber that are bonded with adhesives. It is available in 3- to 5-inch nominal thicknesses and in various standard lengths. The interior surface of the decking is intended to be exposed and is available smooth, grooved, prefinished, or stained.

LAMINATED WOOD DECKING
Figure 3.18

WOOD PRESERVATION

Insects, decay, and fire represent continual threats to wood. Proper seasoning may reduce some of these hazards, but for the most positive protection, other means must be employed.

Insects

The insects most destructive to wood construction are *termites*, which cause extensive damage, particularly in warm, humid climates. Termites eat wood from within, so that a wood member may appear sound, but may actually be eaten away inside.

Preventive measures include proper drainage to minimize moisture, good ventilation, and impervious concrete foundations. Wood members should be adequately separated from the ground or, if not, they should be pressure treated with a preservative.

Other measures include the use of termite shields, metal strips used to prevent the insects from reaching the wood and poisoning the ground adjacent to the building. Because termites attack from within, surface coatings are generally ineffective, and preservatives that completely penetrate the wood should be used instead.

Carpenter ants use wood for shelter rather than for food. Measures used to prevent termite attack are usually effective against carpenter ants as well.

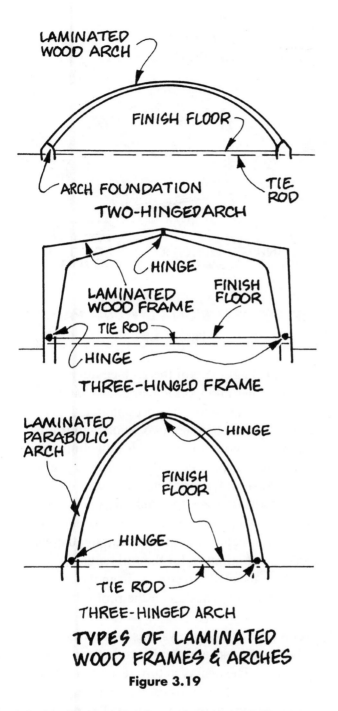

TYPES OF LAMINATED
WOOD FRAMES & ARCHES
Figure 3.19

Marine boring organisms damage wood structures located in salt or brackish waters. The most effective protection is heavy treatment with creosote.

Decay

Decay in wood is caused by fungi that feed on the cell walls. The development of any fungus requires mild temperatures, moisture, and air. Without all three, decay cannot occur. Decay can be avoided if wood is kept dry and well ventilated, or if it is kept continually and completely submerged in water so that air is excluded. Intermittent wetting provides a favorable environment for decay. Where decay is possible, the wood should be a decay-resistant species, such as redwood, or be pressure treated with a preservative.

Preservatives

Chemical preservatives that protect wood from fungi, as well as insects and marine organisms, are of two types: oil-borne solutions, such as creosote; and waterborne solutions, such as chromated zinc chloride. Preservatives may be brushed, sprayed, or dipped, but for maximum effectiveness, they should be applied under pressure for deeper penetration. The oil-borne preservatives cause discoloration or an oily surface that is difficult to paint. Because of the toxicity of preservatives, their use is often restricted or controlled by regulations.

Fire

Fire is an ever-present danger to wood, with the exception of heavy timbers, which burn very slowly. Wood can be made fire resistant by either impregnation with a chemical solution, such as ammonium phosphate, or by the use of a surface treatment. Surface applications, such as intumescent paint, retard the increase of temperature and thereby reduce the rate of flame spread.

WOOD FRAMING

Years ago, when tall trees, and hence large timbers, were plentiful, a designer faced with a 50-foot span would call for 50-foot-long sawn girders to be placed at intervals across the structure. Here was roof framing at its simplest: little engineering, no fabrication, no assembly, and a minimum of small members to be concerned about.

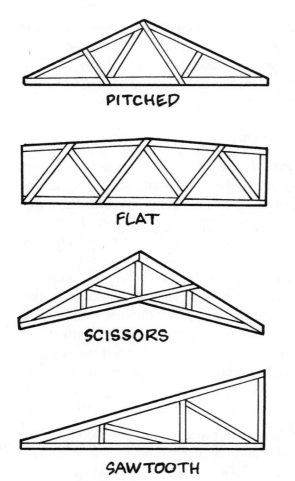

WOOD TRUSSED RAFTER TYPES
Figure 3.20

Today, such large timbers are costly and difficult to obtain. Instead, long-span wood construction utilizes trusses, rigid frames, built-up girders, and glued laminated beams. Several types of trussed rafters (small, closely spaced wood trusses); plywood girders; and a stressed skin panel are shown in Figure 3.20 and 3.21.

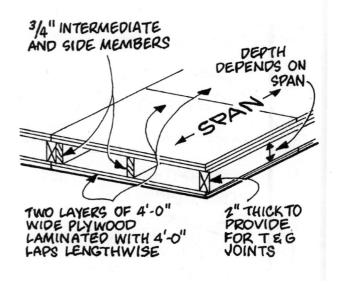

STRESSED SKIN PLYWOOD PANEL
Figure 3.21

Since the beginning of the 20th century, when grading rules and working stresses were established, wood framing has become relatively standardized. There are two wood wall framing systems commonly used for dwellings and other small structures: *platform framing* (also called western framing) and *balloon framing*. In platform framing, the studs are one story in height and the floor joists bear on the top plates of the wall below. In balloon framing, the wall studs are continuous for the full height of the building, usually two stories, from the foundation to the top plates under the roof rafters.

Wood post and beam systems employ beams, rather than joists and rafters, to support floors and roofs. The beams in turn rest on posts or columns, rather than studs, which transfer the building loads to the foundation. The resulting skeleton frame forms a three-dimensional modular grid, which is often left exposed.

Lateral stability is provided by rigid connecting joints, as well as by enclosing exterior walls.

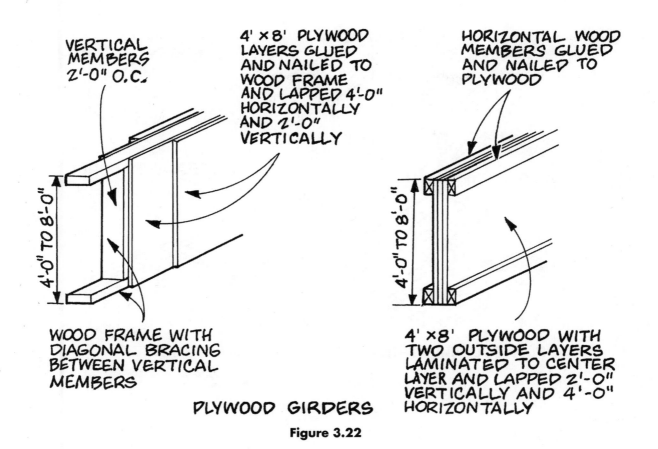

VERTICAL MEMBERS 2'-0" O.C.

4' × 8' PLYWOOD LAYERS GLUED AND NAILED TO WOOD FRAME AND LAPPED 4'-0" HORIZONTALLY AND 2'-0" VERTICALLY

HORIZONTAL WOOD MEMBERS GLUED AND NAILED TO PLYWOOD

4'-0" TO 8'-0"

WOOD FRAME WITH DIAGONAL BRACING BETWEEN VERTICAL MEMBERS

4'-0" TO 8'-0"

4' × 8' PLYWOOD WITH TWO OUTSIDE LAYERS LAMINATED TO CENTER LAYER AND LAPPED 2'-0" VERTICALLY AND 4'-0" HORIZONTALLY

PLYWOOD GIRDERS

Figure 3.22

The extensive subject of wood framing is covered in Chapter 23 of the International Building Code. Candidates are encouraged to review this chapter, with particular attention to nomenclature, stress grades, and common practices used in wood framed construction.

WOOD JOINING

In the days when lumber was plentiful and nails were scarce, it was customary to use large timbers for framing members, and to join them with mortise-and-tenon joints, fastened with wooden pins. The mass production of nails and other fasteners in the 19th century brought about a complete change in all types of wood joining.

Wood joints are classified as exterior or interior, and for both types there is a wide choice of fasteners available, including nails, spikes, screws, bolts, pins, metal connectors, and adhesives. The method of joining is influenced by the species of wood, its characteristics, size, grain, moisture content, exposure, and appearance desired, as well as the necessary strength. Some examples of metal connections are shown in Figure 3.23.

The most common connections in wood structures are made by nails, which are driven into the wood with a hammer or a mechanical gun. *Common nails* are used for most structural connections, *box nails* are thinner than common nails and consequently have less holding power, and *finish nails* are thin nails with small heads used to attach finish wood elements.

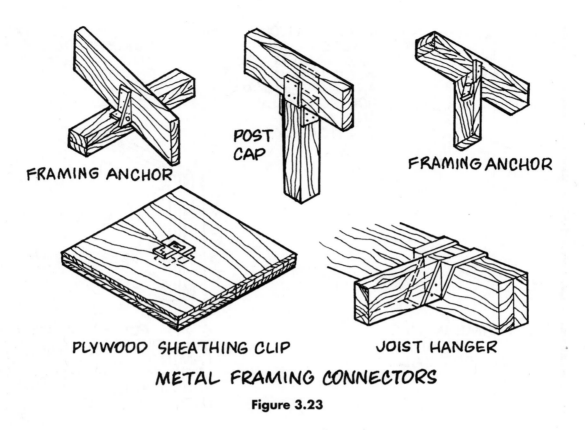

FRAMING ANCHOR

POST CAP

FRAMING ANCHOR

PLYWOOD SHEATHING CLIP

JOIST HANGER

METAL FRAMING CONNECTORS

Figure 3.23

Wood joining is done either at the site or at the mill or shop. Exterior joints must consider weathering, structural joints must meet structural requirements, and interior joints largely consider appearance. In general, the closer a joint approaches invisibility, the more difficult and costly it becomes. Some common wood joints are shown in Figure 3.24.

FINISH WOODWORK

Millwork consists of shop-fabricated items, such as door and window frames, stairways, paneling, etc., which are fabricated in a mill and delivered to the job, ready for installation. Millwork should always be protected by a primer or sealer coat before leaving the shop. Once on the job, it should be handled and stored with care to prevent warping, opening of joints, or other damage. During installation of millwork, ventilation should be provided so that decay or dry rot will not develop where members trap moisture against an unfinished wall.

Wood trim, molding, and ornamental shapes are fabricated from the better grades of both softwood and hardwood. They may also be made from material composed of wood fibers or chips held together with plastic binders and then fused under heat and pressure. All of these shapes are used principally to cover joints where two elements meet and are likely to pull away from one another, or where joints would be unattractive if left exposed. Trim shapes are generally applied by finish carpenters using conventional tools and fasteners. Some common shapes are shown in Figure 3.25.

Wood siding and paneling are made from common dimension lumber, cut and dressed to standard sizes, and available in a variety of shapes and species. Almost all siding is made from seasoned softwoods, while paneling may

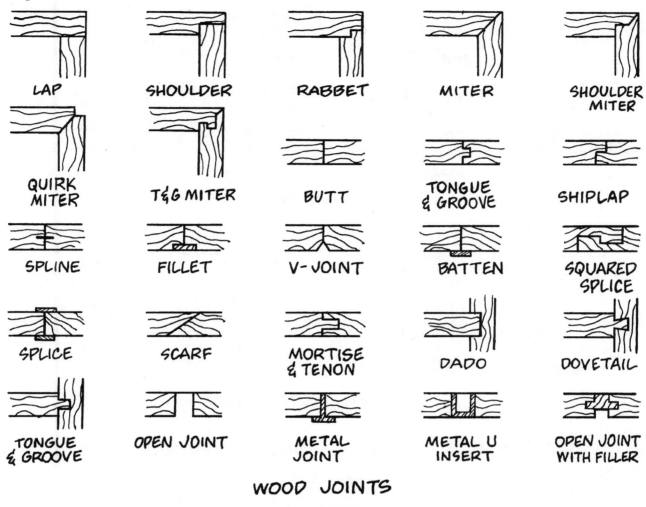

WOOD JOINTS
Figure 3.24

be manufactured from both softwoods and hardwoods. Only the better grades of wood are used for siding and paneling, which are stained, painted, or allowed to weather naturally.

Exterior siding patterns should be selected for their ability to withstand weather, and they should always be applied over lapped building paper. Shown in Figure 3.26 are some common wood siding patterns.

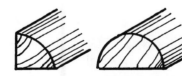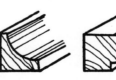

¼ ROUND   ½ ROUND   COVE   BEAD   CROWN   CASING   BASE

WOOD TRIM SHAPES
Figure 3.25

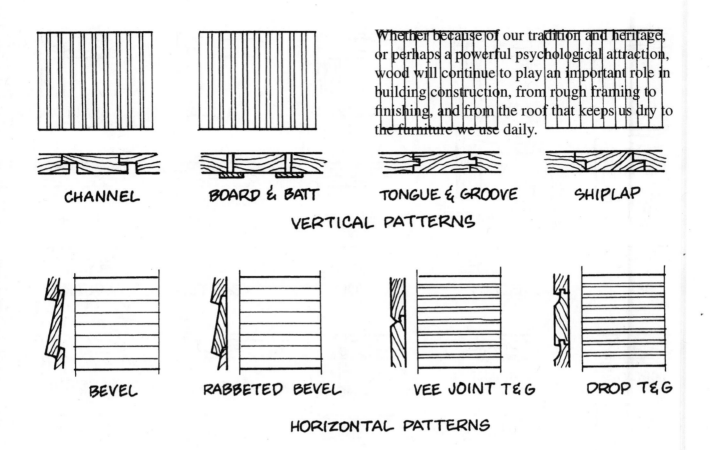

Whether because of our tradition and heritage, or perhaps a powerful psychological attraction, wood will continue to play an important role in building construction, from rough framing to finishing, and from the roof that keeps us dry to the furniture we use daily.

CHANNEL   BOARD & BATT   TONGUE & GROOVE   SHIPLAP

VERTICAL PATTERNS

BEVEL   RABBETED BEVEL   VEE JOINT T&G   DROP T&G

HORIZONTAL PATTERNS

WOOD SIDING PATTERNS

Figure 3.26

CONCLUSION

The future of wood in construction appears to be very promising for several reasons. First of all, reforestation has become widespread, assuring a continuous supply of the raw material. Second, engineered wood products continue to be developed through technology and the imaginative efforts of industry. Finally, we have yet to find another material for small structures that is as inexpensive, easily worked, and universally popular as wood.

END DAY 5

LESSON 3 QUIZ

1. If the primary concern for a structure is fire resistance, one should use

   A. post and beam framing.

   B. heavy timber framing.

   C. platform framing.

   D. balloon framing.

2. Which of the following will help extend the life of wood?

   ✓ **I.** Keep all underfloor wood members dry and well ventilated.

   ✓ **II.** Keep subsurface wood supports totally submerged in water.

   III. Use hardwood members in areas close to the earth.

   IV. Use antitermite chemicals on the surface of wood members close to the earth.

   ✓ **V.** Apply intumescent paint on all exposed wood members.

   A. I and V   **C.** I, II, and V

   B. II, III, and IV   **D.** I, III, IV, and V

3. An architect discovers that the anticipated load of a proposed floor exceeds the safe carrying capacity of conventional wood joists spaced at 16 inches. Because the design is limited to a relatively shallow framing depth, the practical solution would be to use

   A. conventional joists more closely spaced.

   B. shallow wood trusses.

   C. glued laminated beams.

   D. plywood girders.

4. Mortise-and-tenon joints were used in most 18th century timber framing because

   A. it was the easiest type of joint to produce with the available tools.

   B. joint strength could not be achieved by any other method.

   C. metal fasteners were scarce.

   D. appearance was of primary importance.

5. The principal reason that lumber is graded is to establish

   A. uniform standards of quality.

   B. uniform standards of appearance.

   C. the strength of a particular member.

   D. the characteristics of a particular member.

6. Wood that has a moisture content of 30 percent is generally

   A. air dried.   **C.** dry.

   B. kiln dried.   **D.** green.

7. Which type of manufactured panel would be most appropriate to use for perimeter foundation wall insulation?

   A. Beadboard

   B. Fiberboard

   C. Flakeboard

   D. Hardboard

8. Which type of manufactured panel is generally used as core stock for hardwood wall paneling?

   A. Prefinished hardboard

   B. Tempered hardboard

   C. Particleboard

   D. Plywood

9. If you wanted to use wood siding with a highly figured grain pattern, you would specify

   A. rift cut lumber.

   B. plainsawed lumber.

   C. quartersawed lumber.

   D. factory and shop lumber.

10. The strength of a wood member is affected by which of the following?

   I.   The species of tree from which it is cut

   II. The method used in its seasoning

   III. The way it is cut from the tree

   IV. The number of defects it has

   V.   The direction of the applied loads

   A. I, IV, and V

   B. II, III, and IV

   C. I, II, III, and V

   D. All of the above

METALS

INTRODUCTION

All metals come from the earth, usually as metallic ores, which are metal-bearing minerals or <u>rocks</u>. There are a few metals, however, that are found in a purer state, such as <u>gold</u>, <u>silver</u>, and <u>copper</u>.

HISTORY

Discovery and Early Uses

The fashioning of copper into tools, utensils, and ornaments dates back to the Neolithic era. Copper plumbing tubing has been found in Egyptian tombs and palaces from around 3000 BCE. The smelting of copper with small quantities of tin to make brass and bronze appeared all around the world, beginning as early as 3500 BCE, and spurred a sprawling trade network. The Egyptian technique of casting bronze doors for temples was later used throughout the ancient and medieval world. Archaeologists now think that it is the disruption of trade in tin that led to the waning use of bronze and the evolution of iron as the major metal, although bronze is actually much stronger and less brittle than iron.

Iron in its natural state was mined from meteoric rock in ancient times, thus accounting for its rarity. Both the Egyptians and the Mesopotamians called it the "metal from the sky." Carbonized iron, a form of steel used for weaponry, has been made since ancient times across the world. Chinese metallurgy was much more advanced than that of the western world. After the second century BCE, Chinese metallurgists were puddling molten pig iron into wrought iron in a blast furnace, a technology not developed in the West for another 2,000 years. Before the 19th century, the use of iron in construction was limited to tie rods, nails, straps, and decorative applications such as railings, fences, and grills. This iron was either *wrought*, hammered to remove impurities and shape its form, or *cast*, heated and poured into a mold. Those making iron, and its alloy, steel, were not able to control the impurities and carbon content well enough to produce metals with the strength needed for building construction.

Major Developments

Technological developments rendering metals more viable for structural purposes came about during the Industrial Revolution. In 1784, the Englishman Henry Cort patented a dry puddling process for producing wrought iron. Low-carbon wrought iron performs well in tension, and can span greater distances than timber beams. Improvements to the blast furnace process produced a high-carbon cast iron with tremendous compressive strength.

Structural skeletons designed with cast iron piers and wrought iron beams were produced in the 1800s in response to the American impetus to build taller buildings. Cast iron supports, easily manufactured in large quantities, were an inexpensive alternative to masonry. The English gardener Joseph Paxton won a competition to design the exhibition hall (the Crystal Palace in Paris, 1851) for the first World's Fair using a prefabricated iron and glass framing system. Soon iron and glass canopies were being designed to cover shopping arcades, libraries, and train stations.

Iron frame construction was directly replaced with steel after Henry Bessemer, another Englishman, made the production of this stronger metal more efficient and economical with a new production process in 1850s. Shortly after this, another economizing steelmaking process, the Siemens-Martin open-hearth furnace, was developed in Europe. These complementary processes were eventually made obsolete in the 1950s by the process of basic oxygen steelmaking.

Notable Movements, Architects, and Structures

Although the Industrial Revolution of the late 18th century brought revolutionary technology and materials for building construction, the difficulties of life in crowded modern cities revived a nostalgic longing for simpler times and more traditional materials. Beaux Arts architects resisted the use of metal in architecture, leaving it mostly to the realm of civil-engineered structures, warehouses, and factories. They felt that metal was a "scientific" element of architecture incapable of expressing artful forms. They were willing to use metal where it was structurally advantageous, but often hid the metal in a masonry structure. The earliest skyscrapers, such as William Le Baron Jenney's Home Insurance Company Building, built in Chicago in 1885, supported masonry walls on a cast iron and steel frame. The metal piers were encased in masonry, rendering them fireproof and hidden from sight.

The most progressive lines of experimentation were in the design of industrial structures and in temporary structures for the late 19th century world's fairs, where the use of metal was considered appropriate. The Eiffel Tower, intended to be the centerpiece of the world's fair of 1889 in Paris, is a 984-foot tower of wrought iron designed in the shape of a moment diagram to resist the wind. Although Parisians fought the construction of this "iron monstrosity," fearing it would mar the city's skyline and cast deep shadows, it has of course become their most beloved landmark.

Early 20th century architects turned to new materials for a fresh vocabulary to define a modern style of architecture. The Russian Constructivists and the German designers of the Bauhaus envisioned revolutionary structures in steel and glass. The steel and glass skyscrapers built in the United States after World War II, such as Philip Johnson's Seagram Building built in New York in 1958, embodied an image of economic success for new corporations. Although aluminum was produced in the 1800s, it was not used for structural purposes until 1950, when Le Corbusier experimented with framing the glass curtain walls of the United Nations Headquarters in New York in this strong but lightweight metal.

In the 1970s, Frank Gehry tested conventional views of metal as an industrial or commercial material, using chain link and corrugated steel to renovate his own home in Santa Monica, California. He continues to challenge preconceptions about this material by creating voluptuously curving steel forms to enclose Deconstructivist works such as the Guggenheim Museum in Bilbao, Spain.

Recent Trends

Innovations in metals today are in the direction of designed materials made to meet specific construction or design purposes. Special steel alloys can be designed for higher strength, greater plasticity, greater malleability, or lower cost. Weathering steels such as Cor-ten are more environmentally friendly since their pre-rusted surfaces do not need painting. Structural aluminum has the same strength as steel, but its high cost and reduced resistance to heat and deformation makes it less popular. That said, new engineering methods and structural shapes are being developed specific to this material; in the future, structural aluminum may be as widely used as steel and reinforced concrete.

Mining and smelting metals disrupts and pollutes land, rivers, and wildlife habitats. Waste products from these operations can contain significant amounts of lead, arsenic, and other chemicals hazardous to the environment. The application of paint, protective coatings, and fireproofing to metals also causes some pollution. The steel industry has made significant inroads to reduce pollution, but there is still much to be done.

In the United States, the Environmental Protection Agency regulates waste storage and remediation of areas no longer in use. However, there are products within this waste that can be recycled. Most structural steel is composed of recycled scrap; 66 percent of all steel is eventually recycled. Copper building products are fully recyclable; in fact, over half of modern production comes from recycled copper. Recent research has shown that water runoff from copper roofs is safe and environmentally friendly.

The light weight of steel, compared to concrete or masonry, gives steel frames a sustainable advantage. They require smaller foundations than other structures, and steel frame fabrication and erection is a relatively clean process. Aluminum is even lighter and resists corrosion without added finishes, but the higher cost of electricity to produce it outweighs its sustainable advantages.

Iron

Ironmaking improved significantly in the Middle Ages due to experimentation with furnace design. Ironworkers on the Spanish Peninsula invented the Catalan forge around 1293 CE. This open hearth furnace produced a greater quantity of wrought iron with greater speed. Wrought iron rods and chains made with this method were used in the building of masonry domes, such as Brunelleschi's dome for the Florence Cathedral in the Early Renaissance. The Catalan forge was introduced to the Spanish American colonies, where it was improved, became known as the American Bloomery, and was used until the early 20th century. The development of the blast furnace around 1350 resulted in the first pig iron suitable for casting.

In the late 1700s, Henry Cort designed a reverberatory furnace using coal, instead of charcoal, over a sand-lined hearth. The iron extracted from the furnace was shingled (power-hammered) into flat wrought iron, which could then be sent to a rolling mill to be formed into standardized shapes. Although almost 30 percent of the iron was lost in the process, this method continued to be used until Joseph Hall developed a more efficient iron-oxide-lined furnace in 1830. These developments led to the widespread use of iron frame construction.

The first all-metal structure was a cast iron bridge built in the late 18th century to span the Severn River in England. The bridge is still in use today despite its many brittle cracks, because its arched form keeps the majority of the structure in compression. Railroad bridge

designers attempted using cast iron girders and, later, wrought iron trusses, with disastrous results. These structures had to be demolished, as did many cast-iron factories built under the misconception that cast and wrought iron were fireproof materials.

James Bogardus of New York was among the first to use cast iron for architectural facades in commercial and industrial buildings. Familiar classical details traditionally carved in stone could be cheaply reproduced in cast iron and painted a variety of colors. The fashion for forged cast-iron facades faded with the development of the steel frame; however, pressed decorative sheet iron continued to find applications in the early 20th century for interior ceilings and as an exterior wall cladding.

Steel

The cities of Damascus and Toledo, in present-day Syria and Spain, were renowned in the medieval world for producing a high quality steel alloy used for weaponry and cutlery. Their technique may have originated with the crucible method for making Wootz steel, made in ancient India and later in China. The technique for making this steel died out after the 17th century due to the depletion of the mines of special ores it required. An alternative, blister steel, was a method of carburizing, or increasing the carbon content, in wrought iron. Crucible steel was rediscovered by the Englishman Benjamin Huntsman in 1742; however, these forms of steelmaking are now obsolete because they only produce steel in small batches, making steel an expensive commodity.

An abundant and inexpensive supply of steel that could be developed for construction purposes first became available in the 1850s with the introduction of the Bessemer process. An American, William Kelly, was developing a similar process to that of the Englishman Sir

Henry Bessemer, but the Civil War interrupted his progress. The discovery of rich iron ore deposits in the Lake Superior region, combined with the newly imported Bessemer process, started a modern revolution in steelmaking in the U.S. In this process, pig iron produced in blast furnaces is added with manganese, silicon, and small amounts of carbon to a large pear-shaped vessel and heated to 2300°F. Air is pneumatically forced into a vessel of molten iron. As the air rises through the iron it burns out the impurities. Manganese is the essential ingredient that converts the iron into steel. The structural properties of the new alloy are greatly superior to iron.

Around the same time, an open-hearth method for steelmaking was developed by the Siemens brothers in England and Martin brothers in France, which is also used in the United States today. Today, most high-tonnage production of steel is done with the improved basic oxygen process developed in the 1950s.

Aluminum

An aluminum cap was installed on the top of the Washington Monument in 1884; but it would be decades before this expensive material would become affordable for building construction. In 1886, the American company Alcoa developed the first method for smelting aluminum. The first architectural applications were made by Art Deco architects who took an interest in using the material for decorative detailing in the 1920s. The towers of the Chrysler Building and the Empire State Building in New York are noteworthy examples. Between the two world wars, prefabricated aluminum sandwich panels designed for the aviation industry found architectural purposes, and soon aluminum frames were extruded for curtain walls, windows, doors, hardware, and fixtures. The package concept for the aluminum door, hardware, and frame appeared with the very first door manufacturing. Doors were initially made of aluminum sheet or strip inserted into extruded or rolled frames; these sheets were later replaced with glass. After World War II, the popularity of large glass and aluminum storefronts increased.

The aluminum curtain wall industry is growing. It began in 1953 when Alcoa, the world's third largest producer of aluminum, hired Wallace K. Harrison and Max Abramovitz to design its Pittsburgh headquarters with a weather-tight aluminum curtain wall attached to a structural steel frame. Currently, standing seam aluminum sheeting for roofing or siding is finding greater use.

Structural aluminum has a high strength to weight ratio compared to other construction materials. However, its greater tendency towards deformation makes it a more appropriate material where dead loads are the greatest concern in the design of the structure. Recently, structural aluminum has found favor in the construction of large clear-span domes, as can be seen in the current renovation of the United States Botanic Garden conservatory in Washington, D.C.

Copper

Hammered sheets and strips of copper for roofing were first produced commercially around 1500 CE. Copper is the lightest of all roofing materials and was shaped over wood frameworks for cupolas and domes. Around 1750, hot and cold rolling mills were developed to manufacture sheet copper, a technique still used today. Sheet copper was first manufactured in the United States by Paul Revere in 1801.

The first extensive use of copper in 20th century architectural design was by Frank Lloyd Wright. Copper panels ornament the exterior of his only built skyscraper, the Price Tower in

Bartlesville, Oklahoma, symbolizing the leaves of the concrete, tree-like structure. Copper wall panels, fireplaces, ornaments, and furnishings are fashioned inside. The M. H. De Young Museum in San Francisco's Golden Gate Park has been renovated using copper plating. The museum was severely damaged in the Loma Prieta earthquake in 1989; renovations were completed in 2005. Copper plating was used as a lightweight alternative to the original cast-concrete ornament of the building. As the copper oxidizes it will harmonize with the natural setting.

There has been a dramatic increase in the demand for copper in the past 60 years due to technological advances that have made it more affordable and versatile. Copper sheets can now be laminated to a structural core material to be used as wall panels, column covers, and roofs. Roofs can also be covered with copper shingles. American copper mills now offer clear-coated copper that retains its natural color and reflectivity, and pre-patinated copper sheets that eliminate the years-long wait for the green-blue patina to form. The Padre Pio Pilgrimage Church in San Giovanni Rotondo, Italy, completed by the Renzo Piano Building Workshop in 2004, is covered by an immense pre-patinated copper roof.

Metal Finishes

Ferrous metals, such as cast iron, steel, and its alloys will oxidize. The resulting surface rust will continue to deteriorate the metal. Metal coatings were originally designed to inhibit this corrosion. The ancient Romans applied protective coats of melted tin to iron and copper cookware to make its use more palatable. This early version of hot-dip galvanizing continued to be used to protect wrought and cast-iron products into the 19th century.

In 1742, as tin was becoming an expensive commodity, the French chemist Melouin found a way to successfully coat iron sheets in zinc. However, organic acids were found to corrode the zinc, making it poisonous for kitchen utensils. Zinc electroplating was first patented by Charles Hobson in 1805, but it was not until the end of the century that electric generators made this method available commercially. In 1837, William Crawford patented his hot-dip galvanizing process, and similar French applications followed. The numerous coatings and mechanical and chemical finishes developed for iron were transferred and adapted for steel when it overtook iron manufacturing in the middle of the 19th century. The gold rush of 1849 brought corrugated galvanized steel roofing to California and across the Pacific to Australia. By 1850, galvanizing was a major industry in England.

Although copper and aluminum are inherently corrosion resistant, a wide range of finishes have been developed for decorative and protective purposes. Experiments with mechanical finishes, painting, and electroplating for aluminum began in the late 1920s. Most chemical and electrolytic methods are patent-protected. Alcoa purchased British patent rights and opened the first American anodizing plant for aluminum in the 1930s. Today, Japan is the leader in the color-anodizing of aluminum. The anodizing materials, the process, and finished product are certifiably "green."

Structural Steel Construction

Until the late 1700s, iron rolling mills only produced flat and square shapes. In 1783, Henry Cort built a rolling mill with grooved rolls to produce wrought iron shapes. This type of mill was copied for steel billet manufacturing almost 100 years later. Since then, new methods for manufacturing a variety of hot- and cold-rolled shapes have been continuously developed. In the early 20th century,

Bethlehem Steel Corporation in Pennsylvania, formerly an ironworks and mill, produced the first wide-flange structural shapes in the United States. The shapes were produced with a grey rolling mill and were used to create the frames of the first steel skyscrapers. Bethlehem soon became the leading supplier of steel parts to the construction industry.

Prefabricated steel parts were easy to assemble on-site, compared to concrete and masonry construction. For many decades, riveting and welding were the most common methods of connecting prefabricated steel parts. Rivets have since been replaced by less labor-intensive metal bolts. Steel frames still had to be made fireproof, and were originally encased in brick masonry or poured concrete. However, these methods added considerable weight and cost to the steel frame. The attempt to find a lighter weight method of fireproofing led to experiments with plaster applied on a metal lathe. Today, even lighter materials have been developed, such as fireproof gypsum boards (which can also serve as an interior finish), and sprayed-on cementitious mixtures.

Trusses were first fashioned in timber in ancient times and their designs reproduced in iron in the 19th century. The curved steel arch was the preferred form for the first bridges, as well as for Palais des Machines, built for the Paris Universal Exhibition of 1889. Many new truss forms were developed for steel construction, often identified by the name of the engineer who first patented its form. The Warren-type girder, taking the form of a plane truss, was patented by its designer in 1848. The Vierendeel truss, designed in 1896 by the Belgian engineer Arthur Vierendeel, has rect-angular rather than triangulated openings. This special type of truss has been used for many bridges and was the frame design for the Twin Towers of the Word Trade Center in New York.

The Chicago School architects played a key role in the development of the steel frame skyscraper, although these structures were designed to carry heavy masonry walls in Neo-classical and Romanesque designs around the turn of the century. In 1951, Mies van der Rohe designed the Lake Shore Drive Apartments, which eulogized the beauty of the steel I-beam. Most tall buildings used steel frames until World War II. Following the war, a steel short-age forced designers and engineers to design with reinforced concrete frames. The overpro-duction of steel in the 1970s brought prices down and once more steel construction became favored. Chicago's Sears Tower, designed by Bruce Graham and Skidmore, Owings and Merrill in 1973, was the world's tallest tower (1451 feet) for more than 20 years.

The race to build the tallest skyscraper contin-ues, and structural steel is the primary building material. The tallest completed building at the time of this writing is the Tapei 101 in Taiwan, passing the half-kilometer mark (1670 feet). The record should pass to the Burj Dubai in the United Arab Emirates in 2009. In 2013, the Freedom Tower under construction on the site of the former World Trade Center in New York will set a new record as it soars to 1,776 feet.

END DAY 1

CHARACTERISTICS

Metals are substances that are characterized by their luster, opaqueness, hardness, ability to conduct heat and electricity, and their superior ability to resist deformation.

Extracting a metal from its ore is called *smelt-ing*, and invariably this involves some sort of heat treatment. Even after it is extracted, the metal still contains small amounts of impuri-ties, which may be further removed by various refining processes. In some cases minor impurities may enhance the metal by forming an alloy that has special properties, such as increased strength, hardness, or corrosion resistance. Because some metals, such as lead, copper, and iron, are very soft in their pure form, they are often combined with controlled quantities of other substances to form *alloys* when used in building construction.

The physical, chemical, and mechanical properties of metals are so diverse that architects can know only some of the significant characteristics of the metals widely used in construction. This data may include strength, toughness, corrosion resistance, appearance, cost, and methods of forming, joining, and handling.

Metals are classified as either ferrous or non-ferrous. The *ferrous metals* are those containing a substantial proportion of iron, such as stainless steel and galvanized iron; while the *non-ferrous metals* are all the others, such as aluminum, copper, and zinc.

DETERIORATION

Almost all metals deteriorate as a result of exposure to air, water, soil, or other chemical agents. Surface discoloration, or tarnish, is merely unsightly; however, corrosion, such as the rusting of iron, can result in actual failure.

Galvanic action, or *electrolysis*, is a relatively common type of deterioration that occurs when different metals, or alloys, are in contact. If this contact takes place in the presence of an electrolyte, such as moisture, an electrical current will flow from one metal to the other, and in time, one metal will corrode while the other will remain intact. The degree and speed of deterioration depend on the amount of moisture present. Dry air will result in slow action, while

an ocean atmosphere will produce intense and swift deterioration.

The following list of metals is arranged in order of galvanic activity. Each metal can be corroded by all that follow it; for example, lead is corroded by brass, and gold is virtually corrosion-proof. In general, metals far apart on the list should not be placed in contact with each other.

1. Aluminum
2. Zinc
3. Iron and Steel
4. Stainless Steel
5. Tin
6. Lead
7. Brass
8. Copper
9. Bronze
10. Gold

In order to prevent galvanic action, different metals should be isolated from one another, or compatible metals should be used. For example, if aluminum siding is applied with steel nails, the nails should insulated from the siding with neoprene washers, or—better yet—aluminum nails should be used in place of steel nails. If uninsulated steel nails are used, the aluminum around the nail will be gradually eaten away, and the siding may eventually fall off.

Other forms of corrosion may be prevented by changing the chemical composition of the metal or by applying a surface treatment. An example of the former is alloying, such as stainless steel, while an example of a surface treatment is the use of a protective coating, such as galvanizing or anodizing.

FORMING METAL

Forming is the process by which extracted metal is transformed into a useful product that has a finished shape. Products may be either cast or wrought.

Cast products are obtained by pouring molten metal into a mold of the required shape and allowing it to cool.

Wrought products are obtained by forcibly shaping solid metal to a required form by a variety of methods. These methods include hot or cold rolling (bars, sheets, strips, angles, channels, etc.); forging (hammering hot metal); pressing (from sheets); drawing (into wires or tubes); and extruding (forcing a hot mass of metal through an opening that has the shape of the required section).

A particular metal or alloy may be more suited to one method of forming than another. In this regard, one should be familiar with the metal's mechanical properties, such as malleability (ease of hammering); ductility (ease of drawing); toughness (resistance to fracture); and hardness (resistance to abrasion).

FERROUS METALS

Iron

The principal constituent of all ferrous metals is iron, which is the second most abundant metal and fourth most abundant element on earth. All commercial ferrous metals contain some carbon, and small variations in carbon content have an important influence on their properties. Pure iron, which is difficult to obtain, is tough, malleable, easily magnetized, and quick to oxidize.

Wrought iron is almost pure iron with a very low carbon content. It is soft but strong, extremely ductile, easily worked, and relatively resistant to corrosion. It can be forged, bent, or rolled into shape, but it cannot be cast, tempered, or easily welded. Historically, wrought iron was the prime metal used for tension members, such as chains, crane hooks, and anchors. In construction today, however, it is used mainly for ornamental ironwork, grilles, plumbing pipes, and outdoor furniture. Wrought iron is available in pipe, sheet, bar, and bent shapes.

Cast iron is produced by resmelting pig iron with steel scrap. It has a relatively high carbon content (2 percent or more) and is available in a variety of types, each with distinct characteristics (white cast iron, gray cast iron, malleable iron, etc.).

Cast iron has a high compressive but low tensile strength. It is easily cast into almost any shape, but it is generally too hard and brittle to be shaped by hammering, rolling, or pressing. Cast iron surfaces are somewhat rough and uneven, but they have good resistance to corrosion.

Steel

Steel is an alloy of iron that contains no more than 2 percent carbon. Structural steel contains about 0.25 percent carbon, plus traces of various impurities.

There are three principal methods of making steel: the open hearth process, the basic oxygen process, and the electric furnace process. They all involve the removal of unwanted impurities and excess carbon from pig iron and the addition of other elements to produce the desired composition.

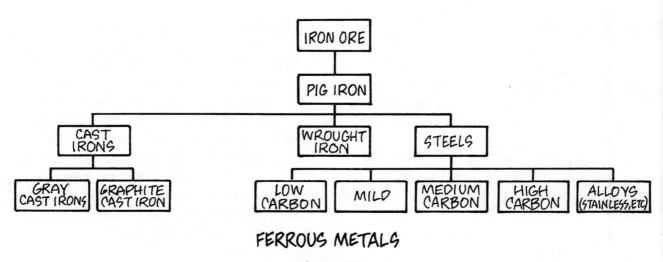

FERROUS METALS

Figure 4.1

Steel is a hard, strong material that is also tough and malleable. It is the most widely used structural metal in building construction, because it provides great strength at relatively low cost. Steel can be rolled, drawn, bent, cast, and joined by rivets, bolts, or welds. It is used for structural framing, concrete reinforcing bars, lathing, conduit, pipes, fixtures, miscellaneous and ornamental work, and for all kinds of connectors, such as nails, pins, and bolts.

Listed below are various types of steel together with some of their properties.

Structural steel. This is steel used for structural purposes, containing varying amounts of carbon and other elements. Included in this type are also a variety of high-strength steels.

Alloy steel. This is steel containing other elements that are added to provide special properties. For example, stainless steel, containing chromium and nickel, is strong, hard, and corrosion-resistant.

Weathering steel. This is steel containing up to one half of 1 percent copper, which develops a tightly-adherent oxide coating when exposed to the weather. It requires no finish.

Heat-treated steel. This is steel that is reheated and cooled slowly, or annealed, for improved workability.

Case-hardened steel. This is steel with a hard, high carbon surface produced by a special process.

NON-FERROUS METALS

Almost all non-ferrous metals share one desirable characteristic: they resist corrosion. In addition, they generally have excellent workability. On the other hand, the initial cost of non-ferrous metals is usually much greater than that of ordinary ferrous metals. As with iron and steel, almost all non-ferrous metals used in construction are alloys.

Aluminum

The important properties of aluminum include its light weight (about one third that of steel); good thermal and electrical conductivity; and great resistance to corrosion (except for galvanic action and some oxidation). It is also highly reflective, making it useful as a barrier to radiant heat transmission.

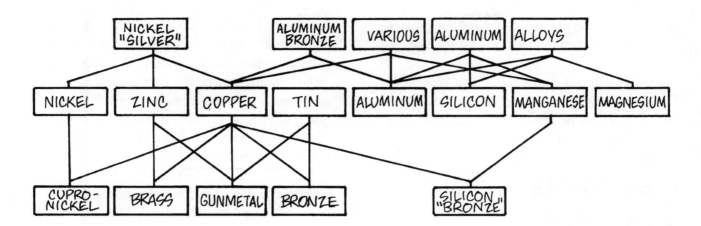

NON-FERROUS METALS AND ALLOYS

Figure 4.2

Pure aluminum is soft, but as an alloy it can be as hard and strong as mild steel. It is available in nearly all fabricated forms, such as castings, extrusions, sheets, strips, bars, and rods. Aluminum may be joined by riveting, welding, soldering, or adhesive bonding. The material takes a great variety of finishes, including etching, embossing, anodizing, plating, baked enameling, and painting (although never with a lead-based paint).

Aluminum is used extensively in construction. Its many applications include framing of lightweight structures, railings and grilles, siding and curtain walls, windows, doors, flashing, insulation, roofing, screening, and hardware.

Copper

Copper is a useful metal that is malleable, ductile, and of fairly high mechanical strength. It is remarkably resistant to corrosive agents, particularly sea water, and it has extremely high electrical and thermal conductivity.

Copper is used in construction for electrical work, water distribution lines, roofing and flashing, and for screening mesh. When exposed to the elements, copper develops a distinctive green patina that makes any further finish unnecessary.

Other Metals

Brass is a common alloy of copper and zinc. In general, all brasses resist corrosion and are easily worked. They are used for precise castings, finish hardware, and for plumbing, heating, and air-conditioning components and fittings.

Bronzes are alloys of copper and tin, with small amounts of other metals. Their properties and uses are very similar to brass.

Lead is a heavy, soft, toxic metal of low strength. It is easily worked, corrosion-resistant, and relatively impenetrable to radiation. It is used in construction for acid and radiation resistance, for vibration control under foundations and machinery, in rough hardware items, and for roofing and flashing (some examples of which are 2,000 years old). However, lead is rarely used these days for plumbing (a term derived from the Latin *plumbum*, or lead), because of the danger of lead poisoning.

Zinc is a relatively low-strength corrosion-resistant metal that is used in building construction for roof covering and flashing and for protective coatings on steel, such as galvanizing.

Monel is a nickel-copper alloy that is strong, bright, ductile, and corrosion-resistant. Its uses in construction include roofing, flashing, counter tops, sinks, and other commercial kitchen equipment.

METAL FINISHES

Finishes are applied to metals for appearance or for protection from corrosion. All metals can be mechanically finished by means of grinding, polishing, sandblasting, hammering, or otherwise treating the surface to obtain a special textural effect. In addition, metals may be treated with an applied coating, such as electroplating, enameling, spraying, dipping, or sherardizing (coating with zinc dust).

Most non-ferrous metals, as well as stainless steels, can be (and often are) left to weather with no additional treatment. Ferrous metals, however, require a protective coating to prevent corrosion. One exception, of course, is weathering steel, which acquires a brownish-colored protective oxide coating that deters further rust and corrosion.

Anodizing

Anodizing is a metal finish applied to aluminum, which begins by inserting the material into an electrolyte. When an electric current is applied, a coating is formed on the aluminum surface in a wide choice of hues.

Galvanizing

Galvanizing is, by far, the most popular method of protecting iron and steel against corrosion. In this process, a coating of zinc is applied by immersing the steel in a bath of molten zinc. The amount of coating is expressed in ounces per square foot of sheet. Even if the zinc coating is broken, because of wear or damage, galvanic action causes the zinc to corrode, thus protecting the iron or steel for as long as any zinc remains. Because of the effective protection it affords at a relatively low price, this use of zinc represents the largest consumption of the metal in the United States.

Galvanized sheets are available in plain flat, corrugated, and special shapes. They are commonly used in construction for roofing, siding, decking, flashing, and cladding, as in kalamein doors (galvanized metal-covered wood core doors used for high fire resistance).

Zinc corrodes to a self-protecting oxide, therefore making it more resistant to corrosive damage than steel. However, zinc is not immune to deterioration. Therefore, when zinc-coated products are exposed to corrosive conditions, they often require the additional protection of paint. Most paints will not adhere to galvanized metal unless the surface has been completely cleaned and prepared. For this purpose a phosphatizing solution may be used to provide the necessary paint bond. Among the available products, portland cement-based paint is an excellent primer for galvanized surfaces.

THE USE OF METALS IN CONSTRUCTION

The use of metals in construction can be classified into six major categories:

1. *Structural*—structural steel, reinforcing bars and mesh for concrete, and wire rope

2. *Hollow metalwork*—doors, bucks, partitions, panels, windows, mullions, curtain walls, and panel systems that incorporate other materials, such as glass, stone, plastic, and so on

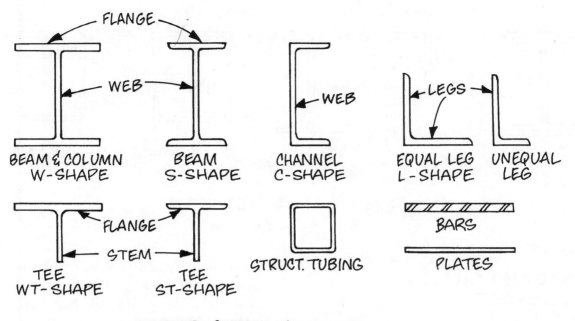

ROLLED STEEL SECTIONS

Figure 4.3

3. *Miscellaneous metalwork*—stairs, railings, fencing, gratings, rough hardware, ladders, and so forth

4. *Ornamental metalwork*—plaques, letters, finish hardware, railings, screens, grilles, expansion joint covers, etc.

5. *Flashing*—base and cap flashing, gutters and leaders, spandrel and through-wall flashing, copings, termite shields, etc. All of these applications are covered further in Lesson Five

6. *Miscellaneous*—rough hardware, nuts and bolts, rivets, screws, nails, washers, inserts, hangers, anchors, wire, and so forth

STRUCTURAL STEEL CONSTRUCTION

Structural steel construction consists of the fabrication and erection of hot-rolled members manufactured from medium carbon steel (about one quarter of 1 percent carbon). Standard

rolled sections are available in a vast selection of weights and sizes, and they include wide flange beams and columns (designated W); American Standard beams (S); American Standard channels (C); angles (L); tees (WT, cut from W shapes); tubing; and bars and plates.

In addition to standard rolled sections, built-up shapes can be constructed from combinations of standard shapes. The resulting configurations, such as plate girders, built-up columns, truss chords, rigid bents, and so forth, are employed where standard rolled sections are not available or efficient.

Structural steel is most often used to construct a skeleton frame. Steel fabrication (cutting, shaping, drilling, welding, and so on) is generally performed in a fabricating shop, where costs and quality can be closely controlled. The work that must be done on the site, such as connecting members by bolting or welding, is kept to a minimum. Structural steel construction requires protection for fire resistance and/or corrosion.

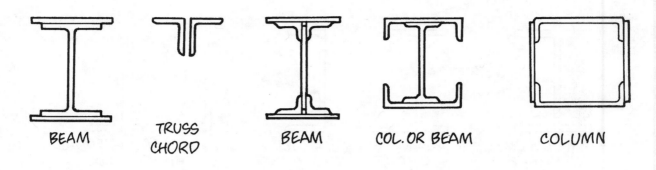

BUILT-UP STEEL SECTIONS

Figure 4.4

TENSILE AND FABRIC STRUCTURES

Tent structures have been around since the earliest times. They were used by nomadic cultures following animal herds, by kings and their courts making the rounds of their territory, and by armies needing portable shelter. The essential feature of the tent is the guy rope, which, held in tension by an upright support and grounded anchors, supports a fabric that creates space underneath.

Tensile structures are designed to have no compression or bending at any point in the system. The first large-scale tensile structure built in iron was designed by the Russian engineer and architect Vladimir Shukhov for the Nizhny Novgorod Fair of 1896. Each of the eight pavilions constructed was composed of a central truss tower holding a lattice of straight iron bars and angle iron in tension. The lattice extended in a radial pattern from the tower to the exterior walls. Shukhov was also one of the first to develop accurate calculations for stress and deformation in tensile structures. Another early example of a membrane-covered tensile structure was the Sidney Myer Music Bowl built in 1958 in Melbourne, Australia. The tensile cable structure is covered with a plywood membrane encased in aluminum.

Since the 1960s, tensile structures have found increasing favor with architects and engineers for the covering of extremely large spaces due to their lightweight, low-cost construction. Frei Otto's 1972 Munich Olympic Stadium, with a seating capacity of 80,000, is the most renowned example. A system of vertical steel supports and cables support sweeping canopies of acrylic glass. Otto also founded the Institute for Lightweight Structures at the University of Stuttgart in 1964.

Recent technological progress in fabric materials for construction has increased the popularity of fabric-roofed structures. Most fabric structures use a woven synthetic or glass fiber cloth that has been coated with a synthetic material to make the fabric airtight and water resistant. The opacity of the material, as well as additional layers, can be used to control heating and cooling and interior acoustics.

Fabric structures are grouped into two types: *tensile* and *pneumatic*. Tensile structures are able to resist wind lift and other stresses due to their anticlastic (saddle-shaped) curvature and by prestressing the fabric and steel cables. A fabric membrane can be held in tension by masts and steel cables or by rigid structural elements such as frames or arches. Tensile loads are transmitted through the structure to ground anchors. Recent designs can be seen at the

Denver International Airport by Leo A. Daly and the Millennium Dome in London by Buro Happold and Richard Rogers.

Pneumatic structures use air pressure to create rigidity in a fabric structure and to resist exterior forces such as wind and snow. This type of structure is most commonly used for sport or recreation facilities, warehouses, or temporary shelters. With air-supported structures, such as the BC Place Stadium in Vancouver, the internal pressurized air supports the prestressed fabric structure above. The double-curvature of the fabric's form is essential for withstanding external loads. Airlocks must be used at entryways to control this air pressure, and the structure must be anchored to the building, foundation, or ground. Air-inflated structures do not require an airlock at the entry. These structures utilize pressurized air between two layers of fabric to give the structure rigidity. Air-inflated structures allow greater thermal and acoustic control.

SPACE FRAMES AND DOMES

The first form of space frame was the tetrahedral box kite, developed by an Australian Lawrence Hargrave in 1893 and by Alexander Graham Bell in 1898. Both men were experimenting with lightweight stable forms for kites and wings that had potential for aeronautic engineering. Buckminster Fuller was the first to use the tetrahedral frame in the design of architecture. The space frame is a relatively new structural form, not widely used until Fuller began building space frames and domes after 1945. Fuller was not the first to develop the geodesic dome, a curved three dimensional space frame, but his work brought it into the public eye. His most memorable work was a geodesic sphere constructed for the Expo 67 held in Montreal, Canada.

TETRAHEDRAL.

The space frame is formed of small, light steel members that together can span great distances with a minimum number of supports, as seen in the Jacob K. Javits Convention Center in New York, designed by I. M. Pei and completed in 1986. A spectacular space frame covers the interior of the Crystal Cathedral, a "megachurch" designed for the community of Orange Grove, California by Philip Johnson in the 1970s. Smaller space frames also can have striking formal qualities, such as the 1989 glass and steel Louvre Pyramid designed by I. M. Pei, or the R. Kemper Crosby Memorial Arena in Kansas City, Missouri. In the Kemper Arena, designed by C. F. Murphy and Associates in 1975, the roof is hung below several space frame bridges, exposing their form on the exterior.

LIGHTWEIGHT METAL FRAMING

Lightweight, or light-gauge, metal framing is analogous to wood framing. Instead of wood studs, joists, and rafters, light-gauge steel members, made of cold-formed sheet steel, are used. Instead of using nails, connections are made with screws, bolts, or welds.

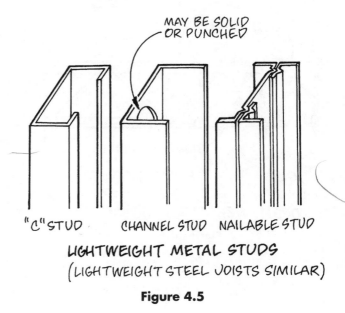

MAY BE SOLID OR PUNCHED

"C" STUD   CHANNEL STUD   NAILABLE STUD

LIGHTWEIGHT METAL STUDS
(LIGHTWEIGHT STEEL JOISTS SIMILAR)

Figure 4.5

Lightweight metal framing involves relatively short spans, up to 32 feet, and relatively light loads. The system has many advantages: it is lightweight, incombustible, and impervious to decay, warpage, shrinkage, and termites. In addition, the components are easily handled and speedily erected. However, lightweight metal framing is usually somewhat more expensive than comparable wood framing.

OPEN WEB JOISTS

Open web steel joists, sometimes called bar joists, are shop-fabricated, standardized lightweight trusses made from hot-rolled or cold-formed steel sections. They are used in a similar fashion to wood joist construction, and they are available in a variety of spans, depths, and load-carrying capacities.

Open web joists may be obtained with a camber, with pre-installed wood nailer strips, or with the top chord forming a conduit raceway. The open web has the advantage of permitting pipes, ducts, and conduits to pass through the joists within the floor construction depth.

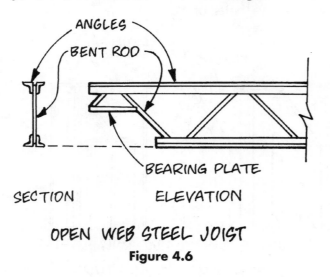

OPEN WEB STEEL JOIST

Figure 4.6

Steel joists are light members and relatively weak in the flat direction. Therefore, they must be carefully handled and stored on the job site.

METAL DECKING

Metal decking is manufactured from sheet steel in a corrugated, ribbed, or cellular form. The edges overlap or interlock to form a working platform during construction and permanent formwork for a concrete floor slab.

The material is either plain or galvanized, with long, narrow sections having ribs about 6 inches on center and 1½ to 4 inches or more in depth.

Cellular steel decking has the additional advantage of providing raceways to accommodate electrical lines.

Most floor decking now is *composite metal decking*, which has deformed or patterned ribs that bond to the concrete so that the deck and concrete act together as a unit to support the loads.

MISCELLANEOUS AND ORNAMENTAL METAL

Ferrous metals are generally used for the multitude of metal items required for a project. These items are often made from conventional fabricated shapes, and they may include metal stairs, railings, fire escapes, gratings, and fences. For this work, shop drawings are usually furnished, in order to check sizes, details, and methods of anchorage.

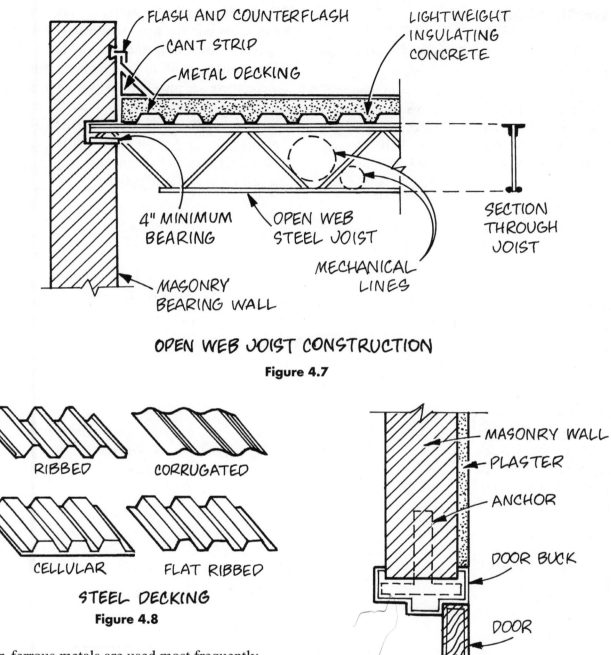

FLASH AND COUNTERFLASH

CANT STRIP

METAL DECKING

LIGHTWEIGHT INSULATING CONCRETE

4" MINIMUM BEARING

OPEN WEB STEEL JOIST

MECHANICAL LINES

SECTION THROUGH JOIST

MASONRY BEARING WALL

OPEN WEB JOIST CONSTRUCTION
Figure 4.7

RIBBED

CORRUGATED

CELLULAR

FLAT RIBBED

STEEL DECKING
Figure 4.8

Non-ferrous metals are used most frequently for ornamental metal work, but stainless steel may also be used. These items are fabricated from rolled, cast, or cold-formed shapes. Among these are decorative grilles and louvers, mesh and wire cloth, metal treillage, and flagpoles.

MASONRY WALL

PLASTER

ANCHOR

DOOR BUCK

DOOR

HOLLOW METAL DOOR BUCK
Figure 4.9

Ornamental expanded metal is made by slitting a sheet of soft metal and then expanding it in a direction normal to the slits.

CONCLUSION

The use of metals in contemporary construction has proliferated as technology has advanced. Today, there are literally thousands of metal items in every building project, regardless of its basic framing material or method. For example, lightweight metal door bucks and hollow metal doors are often used in masonry buildings.

END DAY 3

Similarly, even in the simplest wood-framed structures one can find sheet metal ventilation louvers, aluminum windows, and brass fittings. From curtain walls to flashings to finish hardware, there are metal products used in construction that fulfill the required functions better than any other material is able to do. We can assume, therefore, that metals will continue to be widely used in the construction industry for a vast variety of applications.

LESSON 4 QUIZ

1. All of the following ferrous metals can resist tensile forces *EXCEPT*

   A. alloy steel.

   B. heat-treated steel.

   C. cast iron.

   D. wrought iron.

2. Which of the following metals is most resistant to galvanic deterioration?

   A. Bronze

   B. Brass

   C. Copper

   D. Lead

3. Miscellaneous metalwork is most frequently fabricated from

   A. iron.

   B. aluminum.

   C. copper.

   D. steel.

4. Open web steel joists are

   I.   small lightweight trusses.

   II. standardized.

   III. fabricated in a shop.

   IV. strong both vertically and horizontally.

   A. I and II   C. I, II, and III

   B. II and III   D. I, III, and IV

5. Steel is particularly well suited for structural framing because of its relative

   I.   strength.

   II. ease of fabrication.

   III. low cost.

   IV. corrosion resistance.

   V.   fire resistance.

   A. I and IV   C. II, IV, and V

   B. I and III   D. I, II, III, and V

6. Which of the following is *NOT* a property of aluminum components used in building construction?

   A. High resistance to galvanic activity

   B. Good electrical conductivity

   C. Light weight

   D. Considerable resistance to corrosion

7. Lightweight metal framing systems have all of the advantages below *EXCEPT*

   A. dimensional stability.

   B. speed of construction.

   C. decay resistance.

   D. fire resistance.

8. The purpose of galvanizing iron and steel is to

   A. eliminate all corrosion.

   B. protect against deterioration.

   C. prevent galvanic action.

   D. provide a more attractive finish.

9. Which of the following features apply to metal decking?

   I. It can support normal live and dead loads.

   II. It may be nailed to underlying support members.

   III. Its normal finish affords permanent weather protection.

   IV. It can serve as permanent formwork for a concrete slab.

   V. It is generally corrugated or ribbed.

   A. I and V

   B. I, IV, and V

   C. II, III, and IV

   D. I, II, IV, and V

10. Most fabricated aluminum window assemblies are produced by

    A. casting.

    B. pressing.

    C. cold rolling.

    D. extruding.

THERMAL AND MOISTURE PROTECTION

INTRODUCTION

The earliest building efforts were intended to provide protection and shelter—nothing more. Crude structures protected people from wild animals, unfriendly neighbors, and, not least of all, the weather.

Today, most of the wild animals have disappeared, and unfriendly neighbors are more of an annoyance than a physical threat. The weather, however, is another story; extremes of rain, wind, heat, and cold continue to create problems for architects and those who inhabit their buildings. Temperature extremes cause expansion and contraction of construction materials and, quite often, cracking and deterioration; and excessive heat loss or gain through a structure causes not only discomfort to the inhabitants, but also may require the use of expensive mechanical equipment and energy. Finally, leaks in buildings may range from an annoying, localized drip during a light rain to a damaging flow of water during a heavy storm. The results can be uncomfortable, costly, and—in some cases—actually hazardous.

In recent years, heat loss and gain in structures have become major concerns of building code inspectors, mortgage lenders, and utility companies. The thermal conductivity of exterior walls, roofs, and windows is often regulated in order to minimize excessive heating and cooling equipment and the energy required to run it.

Building users have every right to expect moisture protection and thermal comfort, in much the same way that one assumes the roof will remain overhead. Nonetheless, complete protection of a structure from the elements is often difficult to achieve. It is also a matter that should concern all architects and is the subject of this lesson.

Moisture problems in a structure derive from several sources and exist in various forms. Water is present below ground, at least to some extent, and it may leak into a building by capillary action or because of hydrostatic pressure. Another predictable source of moisture is precipitation in the form of fog, drizzle, hail, rain, sleet, or snow.

Water vapor is present in the air as humidity, and it also exists within a building due to equipment, activities, or simply human habitation. Condensation of this vapor causes water to collect on interior surfaces or within concealed wall, floor, or roof spaces.

GROUNDWATER CONTROL

Surface water consists of ponds and other surface accumulations of water, caused by rain, thawing ice, or snow. This water seeps into the ground to become groundwater.

Groundwater refers to the water contained in the voids and crevices under the earth's surface; this water generally flows very slowly through a permeable material, called an *aquifer*. The level below which the earth is saturated with water is called the *groundwater table*, which can be determined by test borings. Groundwater may have to be controlled or diverted to prevent damage to building materials that it contacts, infiltration into the structure, or footing settlement.

Several terms are used in groundwater control. These are listed in ascending order of watertightness.

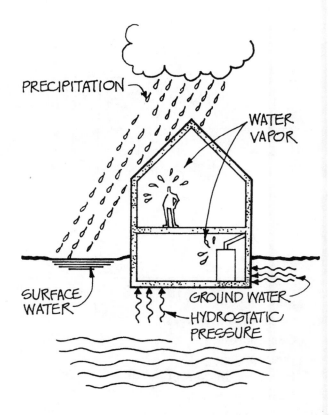

MOISTURE SOURCES
Figure 5.1

Permeable. Capable of being penetrated by water without causing rupture or displacement.

Pervious. Permitting leakage or flow of water through cracks, leaks, or other openings.

Water-resistant. Having no openings larger than capillary pores that permit leakage of water.

Water-repellent. Incapable of transmitting water by capillary action, but able to transmit water under pressure.

Waterproof. Completely impervious to water, whether under pressure or not.

DAMPPROOFING

Dampproofing consists of the materials and methods that prevent moisture from penetrating a building at or below grade. It is distinguished from waterproofing in that dampproofing cannot resist constant hydrostatic pressure. Most dampproofing treatments are applied in liquid form, by brushing or spraying, to the earth-contacting surfaces of foundation or basement walls. These treatments include asphalt base coatings (usually in two coats); cement plaster (densely mixed and troweled or pneumatically applied); and liquid silicones or other plastics.

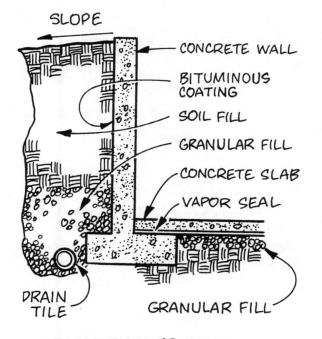

SLOPE

CONCRETE WALL
BITUMINOUS COATING
SOIL FILL
GRANULAR FILL
CONCRETE SLAB
VAPOR SEAL
DRAIN TILE
GRANULAR FILL

DAMPPROOFED WALL
Figure 5.2

Dampproofing may also involve preventive measures, such as draining surface water away from a building; providing granular fill under a concrete slab on grade (to deter capillary action); installing a polyethylene film vapor seal over the granular fill; and employing footing drains (usually perforated clay drain tile) to carry groundwater away from the structure.

WATERPROOFING

Waterproofing consists of the materials and methods that prevent water under hydrostatic pressure from penetrating those parts of a building in direct contact with the earth.

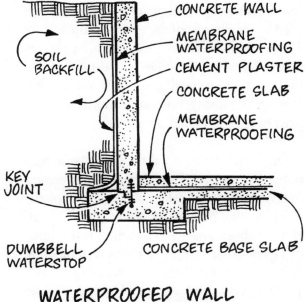

CONCRETE WALL
MEMBRANE WATERPROOFING
CEMENT PLASTER
CONCRETE SLAB
MEMBRANE WATERPROOFING
SOIL BACKFILL
KEY JOINT
DUMBBELL WATERSTOP
CONCRETE BASE SLAB

WATERPROOFED WALL
Figure 5.3

Membrane waterproofing is the most common method of waterproofing subsurface walls. The membrane normally consists of several layers of asphalt-saturated felt (specified as 2-ply or 3-ply) hot mopped together with tar or asphalt pitch. Membranes are applied on the earth-contacting side of walls so that water pressure will force the membrane against the waterproofed surface. Prior to backfilling, wall membranes should be protected from puncture by a coat of cement plaster, a sheet of fiberboard, or a wythe of masonry.

Waterstops are devices used to create waterproof construction joints in walls and floors below grade. Manufactured from noncorrosive metal or various plastics, waterstops permit movement without rupture.

PRECIPITATION CONTROL

Precipitation is the most common and predictable source of moisture that can damage a structure, and its control depends on the form and type of roofing used.

Roofing describes the materials and methods used to weatherproof the exterior top surface of a structure. Several factors determine the selection of a roofing material, including climate, fire resistance, type and slope of roof, weight of roofing, durability, cost, appearance, and—not least of all—personal preference.

Climate affects the durability of roofing materials in numerous ways. For example, strong winds may damage slate or tile, and asphalt shingles may be blown right off the roof deck. Hail has been known to puncture roll roofing and even break thin tiles. Extreme temperatures may cause metal roofs to expand or contract, while asphalt products are adversely affected by high temperatures and ultraviolet radiation from sunlight. Finally, salt, air, smoke, and industrial gases tend to corrode metal roofing, except for those materials resistant to atmospheric corrosion, such as copper, lead, or terneplate.

Fire resistance involves building code classifications A, B, and C, which refer to a roofing material's effectiveness in resisting severe, moderate, and light fire exposures, respectively. These classifications are based on Underwriters' Laboratories designations and have been adopted by most building codes throughout the country.

Type of roof and *slope of roof* are two related factors, as shown in the diagram in the right column. Roof slope is referred to as *incline* or *pitch* and expressed as the ratio between the vertical rise of the roof and its horizontal

projection. For example, a 4:12 pitch is one in which the roof rises 4 inches for every 12 inches of horizontal projection. Roof pitches of 3:12 and greater have little problem in shedding water quickly, while flat or nearly flat roofs depend on a continuous waterproof membrane to contain the water until it drains or evaporates.

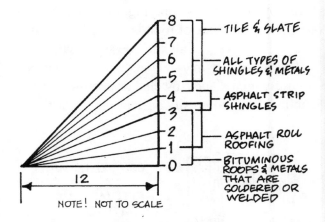

PITCHES FOR VARIOUS ROOFING MATERIALS
Figure 5.4

A *square* is the unit of measure used to express roof surface area and is equal to 100 square feet. For example, a gable roof 30 feet long by 10 feet wide measured along the slope would have an area of 30 feet × 10 feet × 2, or 600 square feet. This roof, then, would require six squares of roofing material (600 ÷ 100 = 6).

The *weight* of roofing materials varies considerably and can affect the design, structure, and cost of the roof. For example, copper roofing weighs about 100 pounds per square, while clay tile may weigh more than ten times as much.

The *durability* of a roofing material is a function not only of its inherent quality, but also the climatic conditions and the manner in which the material is installed. In general, the asphalt products are significantly less durable than the clay, stone, and metal materials.

The *cost* of roofing varies substantially and generally corresponds to its durability, as one might expect.

Personal preference in roofing materials usually relates to *appearance*, and generally speaking, those materials that are the most attractive and preferred also tend to be the most expensive.

END DAY 1

ROOFING MATERIALS

There is a wide range of roof coverings available today, each of which may be appropriately used under specific circumstances. The most common roofing materials, and the forms in which they are available, are as follows:

1. *Asphalt*—built-up, shingles, and roll
2. *Wood*—shingles and shakes
3. *Metal*—sheet, corrugated, and strip
4. *Clay, cement, and slate*—tiles
5. *Glass and plastic*—sheets
6. *Plastic coatings*—liquid coatings

Asphalt

Asphalt and coal-tar pitch are the bituminous materials used in built-up roofing. Asphalt is a by-product of crude oil refining, and coal-tar pitch, or tar, comes from coal. Both products are solid at normal temperatures and are heated to a liquid state before application. Asphalt is generally preferred for sloping roofs and coal-tar for flat roofs. These bituminous materials are used to saturate roofing felts, as an adhesive in built-up roofing, and as a final flood coat in which gravel is embedded.

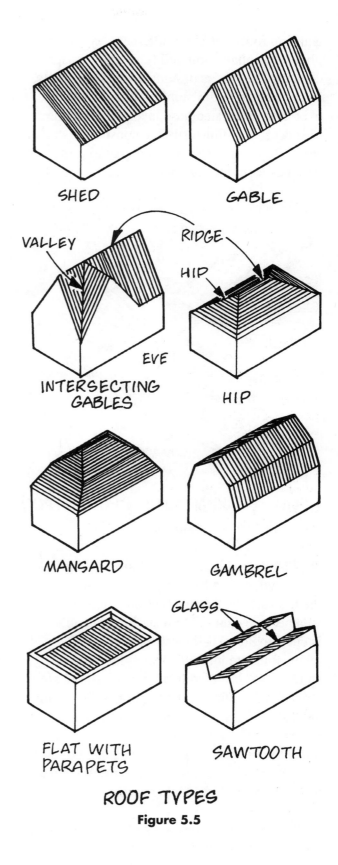

ROOF TYPES

Figure 5.5

Built-up roofs consist of alternate layers (or plies) of asphalt-saturated felts and hot asphalt cement. The roof is designated as 3-ply, 5-ply, etc., referring to the number of layers used over the unsaturated base layer nailed to the roof deck. The finish surface consists of slag or stone chips set in a final flood coat to protect the plies from the weather.

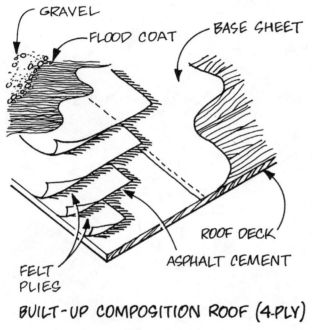

BUILT-UP COMPOSITION ROOF (4-PLY)

Figure 5.6

Asphalt strip shingles (composition shingles) are manufactured from asphalt-saturated felt, the top surface of which has an embedded, weather-resistant mineral surface. The material is cut into 36-inch-long strips that are slotted to simulate individual shingles. Asphalt shingles are installed in a lapped fashion, with galvanized roofing nails, on roof decks having at least a 3:12 pitch. Nearly all asphalt strip shingles are manufactured with adhesive tabs along the underside. These tabs adhere to the shingle strip below, so that the shingles cannot be easily blown off in a high wind.

Asphalt roll roofing consists of material similar to that used in asphalt shingles, but it is manufactured in rolls that are 36 inches wide. It may be used on slightly to steeply pitched roofs, and it is installed by lapping and nailing.

Wood

Wood *shingles* are made from red cedar, cypress, and redwood, because of the decay resistance of these species. They are available in standard sizes, in a variety of grades, and either machine-sawn or hand-split, called *shakes*. Shingles are applied similarly on either roofs or walls in a lapped pattern over solid or spaced sheathing and fastened with aluminum or galvanized nails.

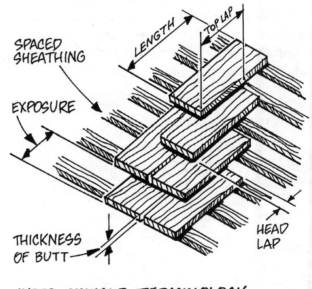

WOOD SHINGLE TERMINOLOGY

Figure 5.7

Wood shingles are tapered in thickness, with the thickness of the butt (the thick end) expressed as the number of shingles required to produce a total thickness. For example, 16-inch 5/2 random shingles means that the total thickness of the butt ends of five shingles, 16 inches long and of random width, is 2 inches. Random shingles (3 to 14 inches in width) are packed by the square, while dimension shingles (5 or 6 inches in width) are packed in 1,000-shingle bundles.

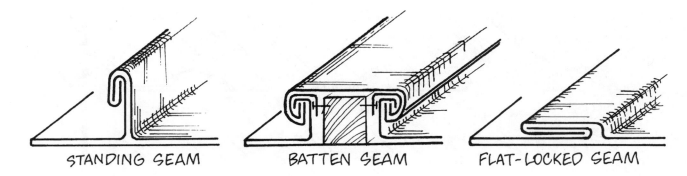

STANDING SEAM   BATTEN SEAM   FLAT-LOCKED SEAM

METAL ROOFING DETAILS

Figure 5.8

Wood shingles are installed on roofs with a slope of 4:12 or greater, and it is preferable to apply them in a way that will permit underside ventilation, in order to prevent cupping or rotting. The use of wood shingles or shakes involves the ever-present danger from fire; however, they may be pressure-treated with fire-retardant chemicals, or sprayed, brushed, or mopped with fire-retardant chemicals after installation. Wood shingle and shake roofs are relatively expensive.

Metal

Metal roofing materials include galvanized iron, copper, aluminum, and terneplate (steel coated with lead and tin). Used to a lesser extent are lead, zinc, and stainless steel. The unique problems involved with metal roofing include corrosion, galvanic action, and movement caused by thermal expansion and contraction. To allow for thermal movement, several installation details are used, such as those shown above. Standing and batten seam roofs create a repetitive visual pattern, which is often used as a design feature.

Corrugated metal sheets, used for both roofing and siding, respond to thermal movement more satisfactorily than sheet metal because of their shape. Filler or closure strips are generally used at the open, corrugated ends.

Metals used on flat roofs are either soldered or welded, while on roofs with a pitch of 3:12 or greater, screws or nails made of a compatible metal are used.

Simulated metal roofing tiles consist of either galvanized steel or aluminum with a baked enamel colored finish. Both types of tile are installed over solid sheathing and roofing felt, using galvanized nails for the steel tiles and aluminum nails for the aluminum tiles.

One of the attractive features of metal roofing is the way it oxidizes when it weathers: lead oxidizes to white, copper turns blue-green, and terneplate weathers to a gray color.

Metal roofs, other than corrugated iron, are relatively expensive, but they are durable and can last for many years.

Clay, Cement, and Slate

Clay roofing tile is manufactured from the same clays that are used for bricks. They are available in a wide range of natural earth colors, as well as glazed finishes, and they are produced in several traditional patterns, as shown in Figure 5.9.

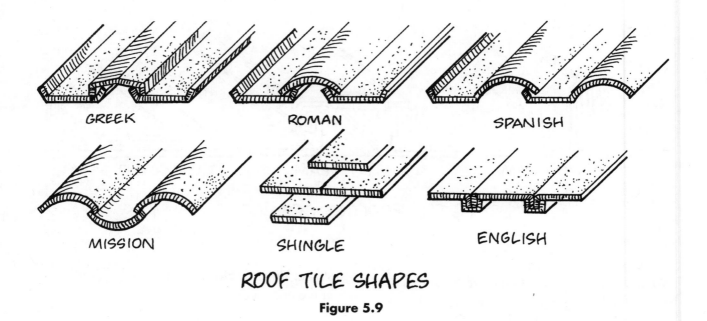

ROOF TILE SHAPES

Figure 5.9

Cement roofing tiles are manufactured from portland cement and fine aggregate. They are generally less expensive than clay tiles, and are available in a variety of sizes, colors, and flat shapes that resemble clay tiles.

Slate roofing tiles (sometimes called shingles) are quarried from natural rock and split into thin rectangular slabs. Available in sizes around 12 inches by 16 inches and from ³⁄₁₆ inch to ½ inch thick, slate is obtainable in a smooth or rough texture and in a variety of natural colors.

All tile roofs, regardless of material, have much in common: they are very heavy, unusually durable and permanent, and they are all fireproof. In addition, all tile roofs are relatively expensive. Tiles are always applied over sloping roof surfaces, lapped over the preceding courses, and fastened with nails applied through predrilled holes.

Glass and Plastic

Glass and translucent plastic sheets are used in skylights, clearstories, and broad roof areas that require light and weather protection, such as industrial structures and greenhouses. Small areas may employ flat sheets, but larger areas generally require corrugated sheets. Both glass and plastic have poor resistance to fire, and therefore wire-reinforced glass and fiberglass-reinforced acrylic sheets are commonly used. Glass used for roofing must be detailed with particular regard for thermal expansion and the low strength of the material.

Plastic Coatings

Plastic coatings are available as single-ply sheets, single-ply with foam insulation, and in a non-solid form ranging from a liquid to a viscous consistency, which can be sprayed, rolled, brushed, mopped, or troweled onto the roof. They are particularly well suited for use on curved, thin-shell concrete surfaces. With all plastic coatings, the substrate must be of a type to which the plastic will permanently bond. Some plastic roofing with thick foam insulation is installed and held in place with heavy gravel or crushed rock.

CONDENSATION CONTROL

Another source of moisture, in addition to groundwater and precipitation, is the condensation of water vapor. Commonly known as humidity, water vapor is always present in the air, to one degree or another; and it is generated within a building by equipment, human breathing and perspiration, cooking and washing, and moisture in the ground that evaporates into crawl spaces.

The higher the air temperature, the more water vapor it can contain. At any given temperature, the ratio of the amount of water vapor the air contains to the maximum amount it could contain is called the *relative humidity*, which is expressed as a percentage. Thus, if air at 50 percent relative humidity is cooled, its relative humidity increases. At some temperature, called the *dew point*, it will reach 100 percent relative humidity and the water vapor will condense into liquid.

Water vapor always moves from high to low pressure areas, usually from inside to outside a building, and it can pass through most building materials. During summer, however, the outside air is often warmer than the air inside a building, and therefore, water vapor may penetrate to the inside of a building. Because condensation can damage almost any portion of a building, water vapor must be prevented from penetrating materials, by using vapor barriers, or be allowed to escape before condensing, through ventilation.

Vapor barriers are materials that prevent the passage of water vapor. They are made from aluminum foil, various types of sheet plastic, or asphalt-saturated felt. Because moisture forms on the cool side of material, *vapor barriers are installed on the warm side*, generally the room side of a wall, beneath the finish material.

Surface condensation of glass can be controlled through the use of insulating glass, and condensation in attics and crawl spaces can be controlled through proper ventilation.

THERMAL CONTROL

Heat is transmitted in three ways: *conduction, convection*, and *radiation*. Conduction occurs when materials or objects are in direct contact. Convection is the process that occurs in a fluid medium, such as air or a liquid. Radiation takes place between two objects not in contact and not shielded from each other.

Heat is gained or lost by a building through these processes, and the purpose of thermal control is to slow down heat gain or loss in order to produce comfortable interior temperatures while conserving energy.

Each element of a building where heat is gained or lost requires a different method of thermal control: *caulking and weatherstripping* are used for cracks and openings; windows utilize *tinted, reflective*, or *insulating glass;* and *thermal insulation* is used in wall and roof assemblies.

Thermal insulation consists of materials that have high *thermal resistance* (R) or a high degree of reflectivity, such as aluminum foil. Some common types of thermal insulation are listed in Table 5.1, along with their R values in square foot-hour-degree Fahrenheit per BTU. Enclosed air has excellent insulation properties, and so air enclosed in cells formed of foamed glass or plastic is often used as an insulator.

The choice of insulation material depends on its physical characteristics, resistance to the flow of heat, and cost. All materials have some resistance to heat flow, but an insulator is arbitrarily considered to be a material whose thermal conductivity (the reciprocal of R, and designated k) for a one-inch thickness is less than 0.5 BTU per hour per square foot per degree Fahrenheit.

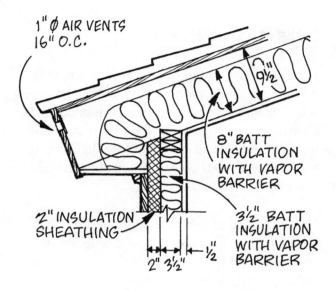

ROOF INSULATION
Figure 5.10

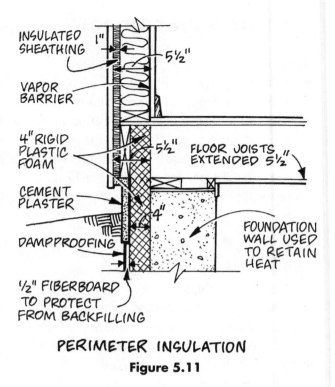

PERIMETER INSULATION
Figure 5.11

Roof insulation can be placed in three different locations: above the roofing membrane, between the membrane and the roof deck, and below the roof deck. Each of these locations requires a different type of insulation, and each has its advantages and disadvantages. There are two advantages of placing the insulation above the membrane: the membrane is protected from temperature extremes and the membrane, being on the warm side of the insulation, acts as a vapor barrier.

Vapor barriers are often combined with insulation, and because condensation tends to form on the cool side of insulating material, aluminum foil–backed insulation should always be installed with the foil facing the warm side.

FLASHING

Wherever different building materials or different parts of a building intersect, there is a natural weakness or gap that may be penetrated by water.

Flashing is material used to provide a seal and prevent water penetration at joints exposed to the weather, intersections of different materials, and expansion or contraction joints.

Flashing is either concealed or exposed. When concealed, flashing may be made from sheet metals, bituminous-coated fabrics, plastic, or other waterproof membrane materials. Exposed flashing is commonly made from aluminum, copper, galvanized steel, zinc, lead, or terne.

Type	Material	Uses	R Value
Loose fill	Glass or mineral wool*, vermiculite, perlite	Wall cavities and flat air spaces such as attics	4" thick = 3.90
Batt or Blanket	Glass or mineral wool* enclosed by paper or aluminum—vapor barriers available	Air spaces in framed walls, floors, and ceilings	3½" thick = 11. 00
Board or Sheet	Cork, glass or mineral fibers, paper pulp	Wall sheathing and rigid roof insulation	1" thick = 2.75
Reflective	Aluminum foil often in combination with layers of paper and air spaces	Roof, wall, and floor insulation plus vapor barrier	1" air space with 2 reflective surfaces = 1.39
Foam	Plastics, spray type, or panels	Sheathing, irregular spaces	1" panel = 6.00

*Rock, slag, or glass, but not asbestos

Table 5.1

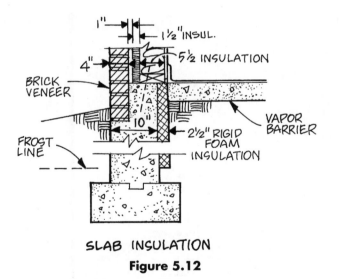

SLAB INSULATION
Figure 5.12

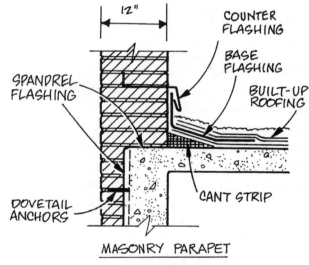

MASONRY PARAPET
Figure 5.13

In selecting a flashing material one should consider durability, possible galvanic action, appearance, and expansion joints where necessary. Flashing is generally included in sheet metal specifications, which may also include scuppers, gutters, downspouts, skylights, and decorative items.

EXPANSION JOINTS

Temperature changes cause expansion and contraction in all building materials, which may produce cracking, distortion, or failure in portions of the structure. The purpose of expansion joints is to anticipate the amount of thermal movement likely to occur and to provide a

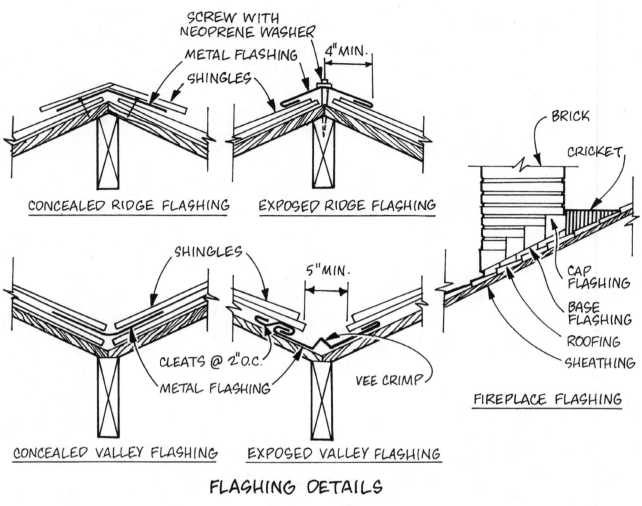

FLASHING DETAILS

Figure 5.14

complete separation that will allow movement, while maintaining the weathertightness and structural integrity of the structure.

Normally, expansion joints in masonry walls are provided every 125 feet or so, and in steel or concrete structures, as well as roofs, joints occur at about 200-foot intervals, depending on the temperature ranges expected. In any case, joints should be located at junctions in L, T, or U-shaped buildings, as well as in stairwells and elevator shafts.

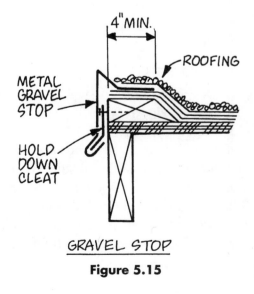

GRAVEL STOP

Figure 5.15

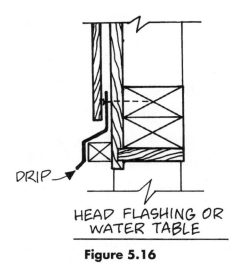

HEAD FLASHING OR WATER TABLE

Figure 5.16

Expansion joints vary from one-half to one inch in width; they provide a complete separation from the top of the footings to the roof, and they are made waterproof with waterstops, elastic joint sealants, metal flashing, or caulking.

END DAY 2.

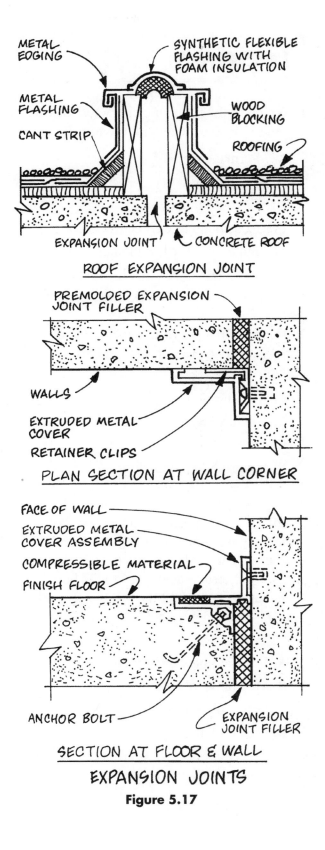

ROOF EXPANSION JOINT

PLAN SECTION AT WALL CORNER

SECTION AT FLOOR & WALL

EXPANSION JOINTS

Figure 5.17

LESSON 5 QUIZ

1. Which of the following characteristics are generally associated with tile roofs?

   I.   Permanence

   II. Low initial cost

   III. Fire resistance

   IV. Considerable weight

   V.   Suitability for low pitches

   A. III and IV   **C.** I, III, and IV

   B. II and V   **D.** I, III, IV, and V

2. The principal purpose of using granular fill beneath concrete slabs on grade is for the control of

   A. condensation.

   B. groundwater.

   C. thermal loss or gain.

   D. expansion or contraction.

3. A sloping roof is framed with 2″ × 10″ rafters at 16 inches on center. The roof decking is ⅝-inch-thick plywood, and the finished ceiling is constructed with ½-inch-thick plasterboard attached to the bottom of the rafters. Which of the following would *NOT* be an appropriate system of thermal insulation?

   A. Mineral fiber rigid sheets

   B. Mineral wool batts

   C. Vermiculite loose fill

   D. Plastic foam panels

4. A polyethylene vapor barrier at roof level should be installed

   A. just above the finish ceiling.

   B. over the insulation on a roof deck.

   C. just beneath the roofing.

   D. between roof rafters.

5. Expansion joints are provided in a structure in order to

   A. reduce thermal movement.

   B. prevent thermal movement.

   C. permit thermal movement.

   D. produce thermal movement.

6. A square of roofing material refers to

   A. 100 square feet of roofing.

   B. the amount of the roof deck area.

   C. the amount of exposed roofing material.

   D. the amount of roofing material, including the necessary overlaps.

7. In which of the following ways can condensation normally be controlled in a structure?

   I.   Employ insulating glass.

   II. Apply a vapor barrier in the walls.

   III. Use a loose-fill type of insulation in the attic.

   IV. Install perimeter drain tiles.

   V.   Provide crawl space ventilation.

   A. II only   **C.** I, III, and V

   B. III and IV   **D.** I, II, and V

8. The preferred minimum roof pitch for the installation of asphalt roll roofing is

   A. dead level. **C.** 3:12.
   B. 1:12. **D.** 4:12.

9. Flashing is generally required to be installed in all of the following situations *EXCEPT*

   A. at the juncture of a masonry parapet and the turned-up composition roofing material.

   B. at the standing seams of a sloping galvanized iron roof.

   C. at the penetration of a sloping asphalt sheet roof by a plumbing vent.

   D. at the head of a metal window frame in a masonry wall.

10. Which of the following insulation materials has the greatest resistance to heat flow per inch of thickness?

    A. Fiberglass batts

    B. Loose vermiculite fill

    C. Aluminum foil with air space

    D. Urethane panels

DOORS, WINDOWS, AND GLASS

Introduction
Doors
Door Hardware
Windows
Door and Window Systems
Glass
Glazing Plastics
Glazing

INTRODUCTION

Wall openings in primitive structures served principally as the means of entering and leaving a building. They also provided light, air, and protection for early dwellers as they watched out for enemies. Holes in roofs were primarily intended for ventilation of open fires, which were used for heating and cooking, but they provided some light as well. No consideration was given to weather resistance, temperature control, or privacy. Thus, these crude openings fulfilled only the basic functional necessities.

In later times, solid doors were developed to provide privacy, weather protection, and defense against intruders, while openings in walls and roofs continued to furnish light and ventilation. In time, window openings were overlaid with frames, which could be opened and closed, and these frames were covered at first with an oiled paper and later with glass. Roof openings continued to be used, but only in warm and arid climates.

Today, doors, windows, and skylights serve a myriad of physical, visual, and psychological functions. Primarily, they allow free access of people, merchandise, light, and air; but in addition, they provide security, privacy, protection from the weather, and access to desirable views. Furthermore, building openings are an important part of the exterior composition; they confirm and often determine a building's scale, and they define the quality and mood of interior spaces.

There exists today a great variety of materials, systems, and products for building openings. Additionally, building codes specify standards for light and air, energy conservation, life safety, and security, to which building openings must conform. Therefore, what was once a relatively simple matter of design is now a complex and unique subject that requires careful study.

DOORS

Doors are the movable barriers that permit or prohibit access to a structure, as well as allow passage between interior spaces. They are held in position by door frames, composed of a *head*, or top piece, and two side pieces, or *jambs*. At the bottom of most exterior door openings are horizontal pieces called *thresholds, sills,* or *saddles*. Almost all doors are manufactured in a mill or factory and installed at the site, or sometimes preassembled, complete with frame and hardware, such as hinges, locksets, weatherstripping, and so on.

Doors are classified in a great variety of ways, such as by location (interior door, entrance doors) or by function (fire door, acoustical door). In addition, doors are classified by their method of operation (swinging door, revolving door); by physical type (paneled door, louvered door); by the material from which they are made (hollow metal door, screen door); and by door hand convention. The following is a description of various door types classified by their method of operation.

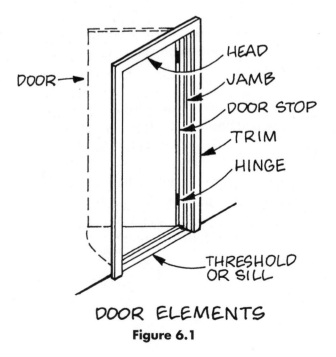

DOOR ELEMENTS
Figure 6.1

Swinging doors are the most common type, and the most effective for weather protection and thermal and acoustical control. They are generally hung on butts or hinges that are attached to the jambs or bucks, but they may also be pivoted from the head and threshold. Swinging doors are most often used singly, but they may also be used in pairs to create a wider opening. When a door is permitted to swing to either side of a jamb, it is called double-acting.

Sliding doors refer to doors that move horizontally. They may roll on a track on the floor or be hung from rollers at the head. In either case, the end of the door opposite the rollers is normally held in place by guides. Sliding doors may operate along a wall surface or slide into a wall pocket.

Folding or *accordion* doors are composed of individual leaves that are hinged together in pairs. They slide along a single guide at the top and often the bottom, and in the closed position, they may form a wall.

For soundproof walls, special seals are used at the top and bottom to prevent sound transmission from one side to the other.

By-passing or *telescoping* doors consist of two or more leaves, each of which travels along its own channel or track. For very large openings, both folding and by-passing doors may be motor-operated.

Overhead doors operate in a variety of ways; they may slide vertically upward, roll up (much like a window shade), fold up in panels, or ride up and pivot inward on special hardware. Generally, all overhead door types are counterweighted for ease of operation. They may be motor-operated, and in the case of residential garages, they are often radio controlled from an individual car or from inside the house.

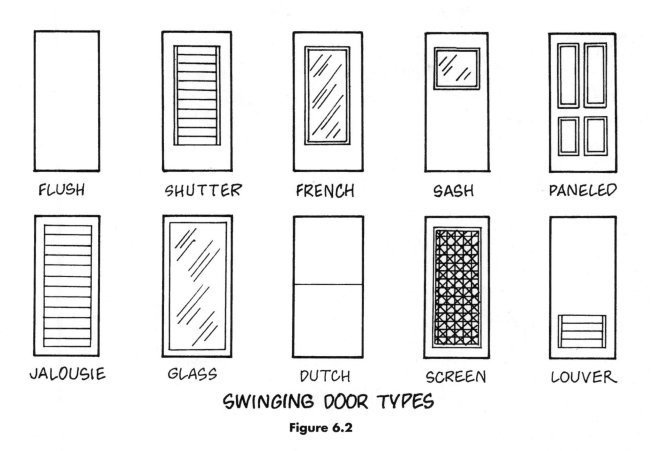

SWINGING DOOR TYPES

Figure 6.2

Revolving doors usually consist of four leaves, at right angles to each other, that rotate about a vertical axis within a cylindrical enclosure. Their principal advantage is the ability to carry a continuous, two-way flow of pedestrian traffic, with a minimal interchange of air between inside and outside. For this reason, revolving doors are frequently used at entrances to large buildings. Where revolving doors are used for building entrances, swinging doors must also be provided nearby to permit the entrance of large objects, to allow for crowd control, and to provide for handicapped and other legal egress.

The classification of physical door types is based on the way in which door panels are made, such as paneled or louvered, and the material from which they are manufactured. Shown in Figure 6.2 are several door types frequently used in construction.

Materials commonly used in the manufacture of doors include wood, steel, aluminum, and glass. A discussion of each of these follows.

Wood doors are the most popular type used in construction. Frames are invariably made from softwoods such as pine, spruce, or fir; cores are constructed from softwood lumber, particle board, corrugated treated paper, plastic foam, or fiberboard; and face veneers may include softwood and hardwood plywoods, particleboard, and plastic laminates. Almost all wood doors are manufactured in mills under controlled conditions. Waterproof adhesives are used on exterior wood doors, while water-resistant adhesives are used for interior doors.

Flush doors may have a core of solid softwood (solid-core door), or a core made up of small pieces of wood arranged in a grid (hollow-core door).

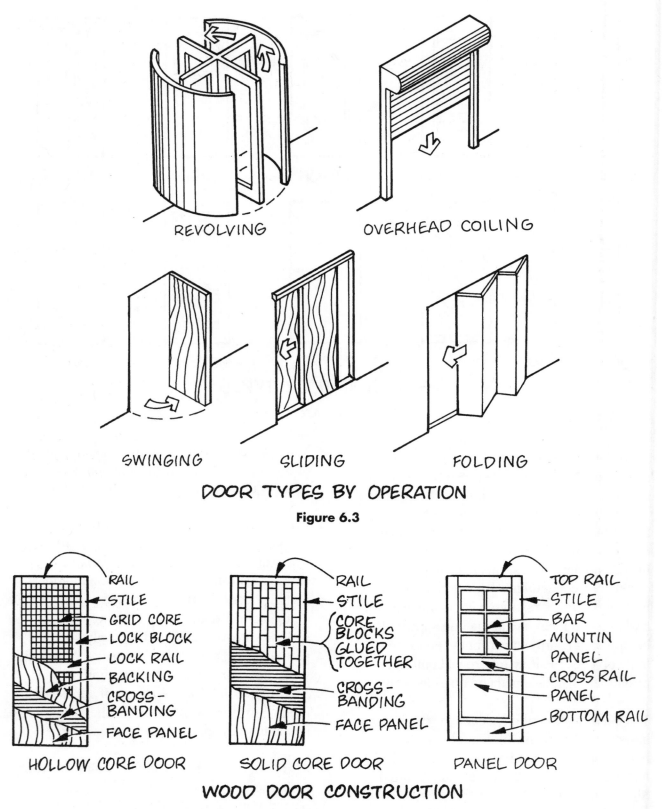

REVOLVING   OVERHEAD COILING

SWINGING   SLIDING   FOLDING

DOOR TYPES BY OPERATION

Figure 6.3

HOLLOW CORE DOOR
- RAIL
- STILE
- GRID CORE
- LOCK BLOCK
- LOCK RAIL
- BACKING
- CROSS-BANDING
- FACE PANEL

SOLID CORE DOOR
- RAIL
- STILE
- CORE BLOCKS GLUED TOGETHER
- CROSS-BANDING
- FACE PANEL

PANEL DOOR
- TOP RAIL
- STILE
- BAR
- MUNTIN
- PANEL
- CROSS RAIL
- PANEL
- BOTTOM RAIL

WOOD DOOR CONSTRUCTION

Figure 6.4

Panel doors consist of various sections held in place by solid wood stiles and rails (vertical and horizontal framework). Panels may be made from solid wood, plywood, glass sections, or fixed wood louvers.

Special-purpose doors are generally wood flush doors manufactured with a special core to resist fire (up to a 1½-hour U.L. approved rating), sound, or radiation.

Steel doors are generally used for interior installations that require fire ratings and for exterior doors with or without fire ratings. They are available as swinging doors, and also as roll-up, fold-up, sliding, folding, bypassing, revolving, etc. Except for cast metal doors, which are used rarely and only for monumental purposes, steel doors are usually manufactured from 14 to 20-gauge cold-rolled steel, or in some cases, stainless steel, which requires little maintenance.

Hollow metal doors are constructed of a steel frame covered with sheet metal, and they are available either flush or paneled. The metal faces are stiffened by the use of steel reinforcing shapes or by solid core material, including particleboard, rigid foam, or other types of material. The resulting doors are rigid, permanent, and can be fabricated to meet any fire rating required by building codes.

Metal clad doors, commonly called *kalamein* doors, consist of a solid wood core covered with sheet metal. The core will ordinarily resist fire as long as the sheet metal cover prevents oxygen from reaching it. In recent years, kalamein doors have largely been replaced by fire-rated, solid-core doors.

Aluminum doors are normally fabricated from extrusions or rolled shapes. They are lightweight, corrosion-resistant, and dimensionally stable. However, aluminum doors must be protected from staining, scratching, and galvanic action. Aluminum doors are most commonly used as entrances, for storefront work, and in curtain wall construction. Aluminum doors are rarely used for fire doors because of their low melting point. However, there are special aluminum fire doors that are constructed with a fireproof core.

Metal security doors are composed of a steel frame covered by metal mesh or grilles. They operate in a variety of ways and serve to protect commercial and industrial buildings from unlawful entry.

Fire doors are used in fire-resistive construction and are U.L. rated and labeled, from class A to E, according to the door's location. Table 6.1 describes the various classifications. In addition to the requirements in the table, fire door assemblies (frames, hardware, etc.) must have the same rating as the labeled door. The maximum fire door size is 4 feet by 10 feet, and all doors must be self-latching and equipped with automatic closing devices.

Glass doors are made of ½ to ¾ inch thick transparent or obscure tempered glass and used for swinging or revolving doors. After tempering, the glass cannot be cut, drilled, or altered in any way. Glass doors are generally pivoted at the top and bottom, as it is difficult to support their great weight with conventional hinges.

END DAY 1

DOOR HARDWARE

Door hardware comprises the largest portion of the finish hardware specifications. It includes all items that permit doors to operate properly, such as hinges, closers, locking devices, panic hardware, and weatherstripping. The selection of door hardware is governed by the

Door Label	Fire Rating in Hours	Location in Structure	Glazing Permitted (¼" wire glass)
A	3	Fire walls separating buildings or fire areas within a building	None
B	1½	Vertical enclosures— fire stairs and elevators	100 sq. in./leaf (4" min. dimension)
C	¾	Corridors and partitions	1296 sq. in./light (54" max. dimension)
D	1½	Exterior walls— severe fire exposure	None
E	¾	Exterior walls— moderate fire exposure	720 sq. in./light (54" max. dimension)

Table 6.1

required function, operating characteristics of the devices, durability, weather resistance, and appearance.

Originally, door hardware was always metallic, such as iron, steel, and stainless steel for heavy-duty installations, and aluminum, brass, and bronze for other uses. Now, however, ceramics and plastics are increasingly used. The location of door hardware has been established through years of usage and varies only slightly. Thus, door knobs are invariably located around 38 inches from the floor, and door pulls, as well as panic bolts, are placed around 42 inches from the floor.

Door hand convention is a standardized uniform reference method used by architects and hardware manufactures and suppliers. The hand of door is determined in the following way: Assume that you are standing outside the door and facing the door. In this case, the hinges are not visible to you, and the door swings *away* from you. If the hinges are on the left side, the door is left-hand; if the hinges are on the right side, the door is right-hand. If the hinges are visible and the door swings *toward* you, the door is either left-hand reverse or right-hand reverse, depending on whether the hinges are

on the left or on the right. Figure 6.5 should clarify door hand convention.

Hinges are the devices on which doors swing or pivot to open and close. They may be exposed, concealed, or invisible; but most commonly they are mortised into the door edge and jamb, so that only the knuckle is visible when the door is closed. Because only the butt end of the hinge is exposed, hinges applied to the edge of a door are known as *butts*, or *butt hinges*. Most hinges are made up of three parts: two leaves, one with an odd and the other with an even number of knuckles, and a pin that joins the leaves. The pin may be removable (loose) or nonremovable (fixed or non-rising) for security situations. Pin tips are commonly oval or flat, but hospital tips are curved for ease of cleaning and to avoid catching articles of clothing. Hinges with a standardized screw hole size and pattern are called template hinges.

The size of butt hinges varies from 2½ to 6 inches in length, which in general is the same dimension as the width. With a normal-height door, hinges are mounted approximately 8 inches from the head and 10 inches from the floor. When a third hinge is required, it is mounted midway between the top and bottom hinges. Figure 6.6 shows several types of hinges commonly used in construction.

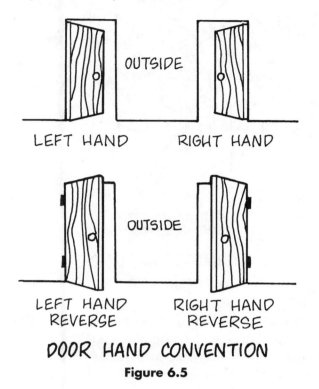

DOOR HAND CONVENTION

Figure 6.5

Closers are devices designed to automatically close doors quickly, quietly, and with no damage to the door or frame. In addition, they may also be used to hold a door open. They are either exposed or concealed, and generally consist of a spring (the closing element) and some type of hydraulic mechanism. Closer installations may be of the parallel-arm type (minimum projection), or bracket-mounted type. Door closers are selected on the basis of door type and size, frequency of operation, exposure to weather, and appearance. A fusible link closer is used to close a door automatically in case of fire.

Locking devices are used to hold doors in a closed position and to provide security. When the locking mechanism is beveled, the device is referred to as a *latch*, and a latch automatically slides into position when the door is closed. When the locking mechanism is rectangular in shape and it must be projected manually, the device is referred to as a *dead bolt*. Such a bolt is often used in conjunction with a latch, in which case the unit is known as a *lock*.

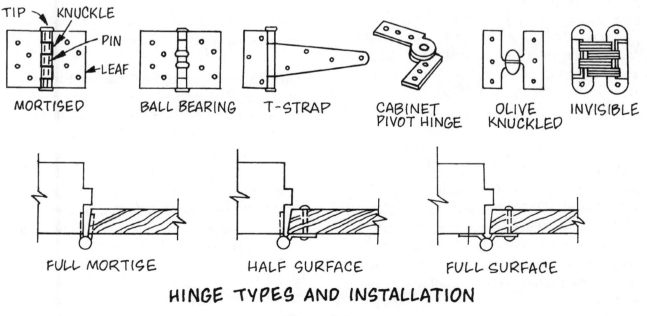

HINGE TYPES AND INSTALLATION

Figure 6.6

Locksets are differentiated by the way they are installed. *Rim locks* are mounted on the interior surface of the door and have square or rectangular boxes. *Cylinder locks* are more or less concealed by fitting into holes drilled in the lock stile of the door. They are popular, inexpensive, and easy to install. *Mortise locks* fit completely into a rectangular cavity carved in the edge of the door. As they are almost entirely concealed, they are the most secure. *Unit locks* fit into a door cutout and are therefore installed quickly and efficiently.

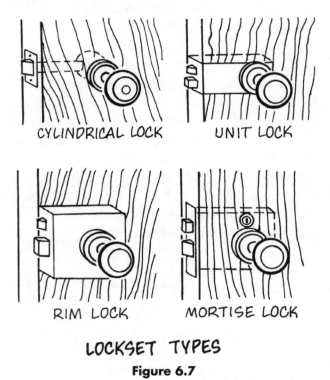

LOCKSET TYPES

Figure 6.7

Panic hardware is used on exit doors of public buildings to quickly and easily open the doors in case of emergency. It is required by building codes, which specify that push bars to release the locking mechanisms must extend across at least three-fourths of the door width. Panic hardware may be of the vertical rod type, which exposes the latching at head and floor; or the mortise type device, which conceals the lateral latching.

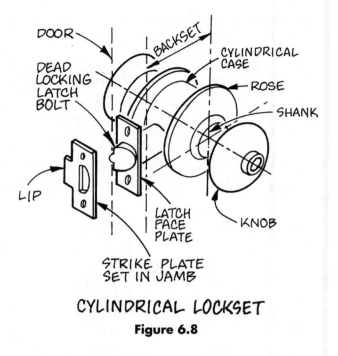

CYLINDRICAL LOCKSET

Figure 6.8

Operating devices for doors include door lever handles, door pulls, push plates, kick plates, and escutcheons (decorative plates covering the locking mechanism). Door knobs should be avoided. Accessibility requirements for public spaces prohibit them, and while they are still allowed for private residences, homes too are better served by levers that provide greater access to the elderly and persons with disabilities.

Operating devices are available in a variety of shapes, designs, and materials. Finishes for door hardware should be selected on the basis of durability, weather exposure, corrosion resistance, and general appearance. Some common finish designations are listed below. The BHMA numbers listed are found in ANSI/ BHMA A156.1, the standard for builder's hardware materials and finishes. The old U.S. numbers are listed parenthetically.

BHMA 600 – Steel primed for painting (USP)

BHMA 605 – Bright brass, clear coated (US3)

BHMA 606 – Satin brass, clear coated (US4)

BHMA 611 – Bright bronze, clear coated (US9)

BHMA 612 – Satin bronze, clear coated (US10)

BHMA 613 – Dark oxidized satin bronze, oil rubbed (US10B)

BHMA 618 – Bright nickel plating on brass or bronze, clear coated (US14)

BHMA 619 – Satin nickel plating on brass or bronze, clear coated (US15)

BHMA 622 – Flat black coated brass or bronze (US19)

BHMA 625 – Bright chromium plated over nickel on brass or bronze (US26)

BHMA 626 – Satin chromium plated over nickel on brass or bronze (US26D)

BHMA 628 – Satin aluminum, clear anodized (US28)

BHMA 629 – Polished stainless steel, 300 Series (US32)

BHMA 630 – Satin stainless steel, 300 Series (US32D)

The BHMA numbers differ from the old US numbers in that they provide not just the visible finish but also the base metal. Thus, where US10B described an oil-rubbed bronze finish, the base metal could have been either solid bronze or bronze-plated steel. Although bronze-plated steel might be acceptable for decorative cabinetry, it would be inappropriate for heavily used exterior commercial building entry pulls.

Weatherstripping is the means employed to make exterior openings weather tight. It acts as a barrier to prevent wind and moisture from entering a building through cracks between doors or windows and their frames. This is accomplished by means of interlocking or friction devices which are manufactured from metals, such as aluminum, bronze, stainless steel, etc. or pliable materials, such as felt, rubber,

and plastic. Some examples of weatherstripping are shown in Figure 6.9.

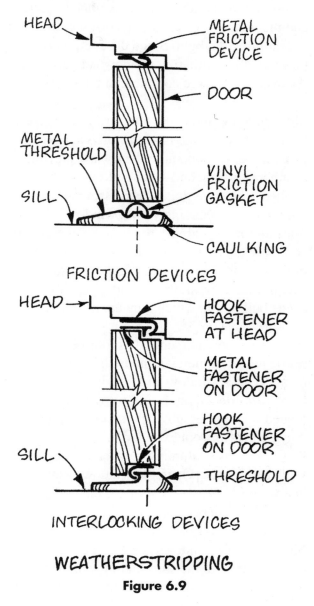

FRICTION DEVICES

INTERLOCKING DEVICES

WEATHERSTRIPPING
Figure 6.9

WINDOWS

Windows are glazed openings constructed in a wall to admit light, air, and, in solar designed buildings, solar heat. The earliest windows were used to provide natural light only, but with the development of hinges and fastening devices, the window also became a means for ventilation. For a number of years after the introduction of air conditioning, natural ventilation in large buildings became unnecessary, and the window returned to its original function of admitting daylight. Since the energy crisis, however, the ventilating window is once again reappearing in large commercial buildings.

Most windows today are standardized in size (stock windows) and manufactured as complete units, with operating hardware, glazing, weatherstripping, and—quite often—a factory-applied protective finish. Aside from admitting light, air, and desirable views, windows are selected for their ability to resist weather, open and close efficiently, provide thermal insulation, and admit solar heat. Windows also dominate the scale, proportions, and general design of the building in which they are installed.

Windows are manufactured in four basic types: residential, commercial, industrial, and monumental. Stock window units are manufactured from kiln-dried, straight-grain wood, which is treated with preservative, paint, or clad with vinyl, and also from extruded or rolled metal sections, such as aluminum, steel, stainless steel, and bronze. Operable windows function in a variety of ways, most of which are illustrated in Figure 6.11.

Most operable windows consist of a frame, which is fixed to the surrounding wall or other supports, and a sash, which is the glazed panel fitting into the window frame. The various parts of a window are shown and identified in Figure 6.10.

Wood windows are manufactured according to well-established standards from woods that are able to resist shrinkage and warping due to weather exposure. The most common of these are ponderosa pine, white pine, sugar pine, fir, redwood, cedar, and cypress. Wood windows

are fabricated in a mill from kiln-dried lumber (6 to 12 percent moisture content), and treated to resist moisture, mold, fungi, and insects. They are also available with a factory-applied plastic finish, or sheathed in aluminum. Both of these finishes eliminate the need for periodic painting or staining. Wood windows are relatively inexpensive, durable, and widely available in a variety of types and standard sizes; however, fire-resistive code requirements sometimes preclude their use.

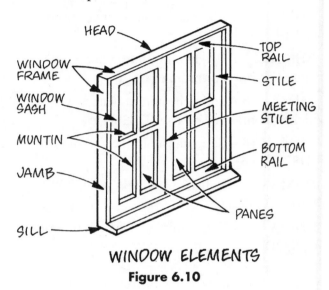

WINDOW ELEMENTS
Figure 6.10

Aluminum windows are made from extruded aluminum alloy sections that are strong, durable, and very light in weight. These aluminum extrusions are particularly well suited for windows, as the complex configurations, which are simple to extrude, can provide for glazing, weatherstripping, and condensation, all within one section. Aluminum windows are available either in a matte or highly polished metal finish, in various colored anodized finishes, and in plastic and baked enamel finishes, both of which protect the metal from oxidation. When installing aluminum windows, all trim, fasteners, etc., must be of a compatible metal or isolated from the aluminum in order to prevent galvanic action. A variety of sizes and types are standard and readily available.

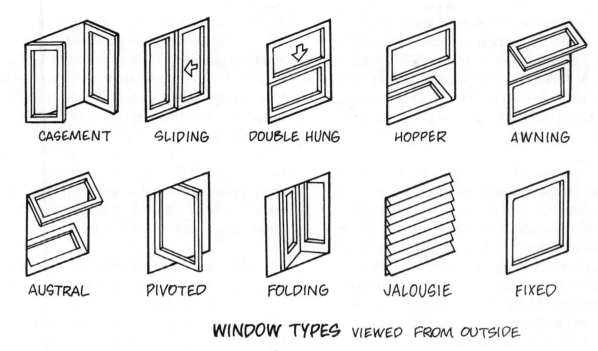

WINDOW TYPES VIEWED FROM OUTSIDE

Figure 6.11

Steel windows are manufactured from hot-rolled solid sections or from cold-rolled strip steel, which is more widely used. Standards have been established for types, sizes, hardware, and operating mechanisms, and most manufacturers conform to these standards. The durability of steel windows depends on their protection against corrosion, which is generally achieved by a factory-applied baked enamel finish or a plastic finish. Frequently, steel windows are glazed on the site.

Stainless steel windows are strong, corrosion-resistant, and usually manufactured from 16, 18, 20, or 22-gauge stainless steel sheet. All operating closers, fasteners, anchors, etc. must be compatible, so that no galvanic action can occur.

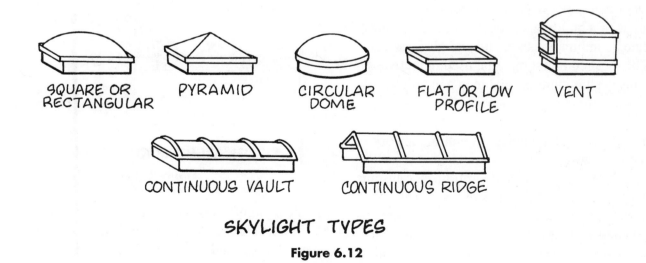

SKYLIGHT TYPES

Figure 6.12

Replaceable windows supersede existing windows, or are applied over an existing window. They are available in a wide range of types, sizes, and materials, and are widely used in alteration and renovation work. They are also used to decrease heat loss and gain in existing buildings.

Screens consist of wire mesh supported by rigid frames placed over window and door openings to exclude insects. The mesh or wire cloth, sometimes called insect screening, is available in a variety of materials and finishes, such as steel, aluminum, copper, plastic, etc., which are painted, galvanized, or otherwise coated to improve durability. The most common insect screening is 22-gauge wire with 12 openings to the inch. Screen frames that support the mesh are fabricated from wood or metal and are available in stock or custom sizes.

A special type of screening is made of sheet aluminum that has been slit and formed into tiny louvers. This type of screening not only excludes insects, but it also controls the infiltration of light and heat.

END DAY 2

DOOR AND WINDOW SYSTEMS

Door and window systems consist of large areas of a building's skin, most often prefabricated combinations of solid and glass panels. The most common material used in these systems is aluminum, available in a great variety of standard extruded shapes.

Storefront work comprises metal framework sections and glass panels designed as a unit. It includes the entrance door, frames, transoms, hardware, and fixed panels of glass. The advantage of this kind of system is the integration and standardization of all elements, that are related in design, function, and installation.

Curtain walls are exterior walls of a building that support no weight other than their own, and function solely as an enclosing skin. They usually consist of metal framework in combination with glass, plastic, metal, or other surfacing panels. The essential functions of a curtain wall are much the same as any other building wall: it must be weatherproof, it must provide some degree of thermal and sound control, and it must be safe and secure. Curtain walls are also considered for their durability, ease of maintenance, and aesthetic suitability.

Skylights are overhead sources of natural light generally installed on a roof. They may be flat, pitched, vaulted (curved), or dome-shaped. Skylight frames are usually fabricated from metals such as aluminum, galvanized steel, or copper. They are glazed with glass block; flat or corrugated wire glass; or clear, tinted, or translucent sheet, corrugated, or sandwich-panel plastics. Prefabricated stock units, glazed with acrylic, are particularly popular in construction.

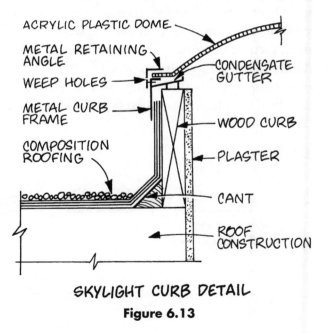

SKYLIGHT CURB DETAIL
Figure 6.13

All skylights must be installed in a weathertight fashion, they must be safe from breaking and falling, and they must be provided with condensation drains to conduct interior moisture to the outside through weep holes in the curb frame. Fire safety is also a concern with the use of skylights, and therefore, their size, position, and spacing are often regulated by building codes.

GLASS

Glass is one of the oldest building materials still used. Some 3,000 years ago it was utilized for decorative purposes in Western Asia, and in ancient Rome glass was first employed in windows. Until relatively modern times, glass was difficult to make and therefore very costly. In fact, during the 18th century, individual glass windows served as a basis for taxation. Today's glass, however, is widely available, inexpensive, and vastly improved over earlier products. Float glass, for example, which was first developed about 40 years ago and is made by pouring molten glass onto the surface of molten tin, produces a glass that is perfectly flat and brilliantly clear. The widespread use of glass has become closely identified with modern architecture, and we can expect that glass of all types will continue to be developed and used.

Glass is a solid, supercooled, ceramic material made from sand, soda, and lime. It is a chemically inert, generally transparent, hard, brittle material that—unlike other ceramics—is shaped in a liquid state, at high temperatures, and then allowed to cool. If necessary, glass can be reheated and shaped again.

Glass is easily worked and can be fabricated in a wide variety of shapes. It can be cast, pressed, rolled, blown, extruded, spun, and ground. Spun glass fibers (glass wool) are used for acoustical and thermal insulation. Glass fibers are also used in textile manufacturing, and alkali resistant (AR) glass fibers can be used as reinforcing in concrete and plaster.

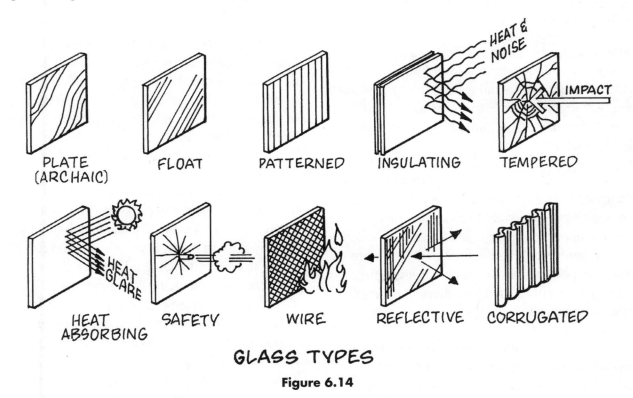

GLASS TYPES

Figure 6.14

The glass industry still uses terms from the more archaic glass processes, such as *plate glass* and *annealed glass*. Presently there are no major plate manufacturers in the United States, and the only true plate glass still used is either recycled (reclaimed from existing applications) or wired glass imported from Japan or Europe. The term *double-strength glass* (DS) might still be used by some to refer to ⅛-inch-thick float glass, but like the old *single-strength* (SS) ³⁄₃₂-inch glass, it should be considered archaic as well—unless specifically referred to by the manufacturer.

Float glass, the typical glass used today, is annealed in the manufacturing process, so there is no need to refer to *annealed glass,* as all float glass manufactured in the United States is annealed. In general, float glass is available from ⅛-inch thick (3mm) to ⅞-inch (22mm) thick. Most glass, even in the United States, is manufactured in metric thicknesses, so the thicknesses should be considered nominal, with standard nominal ¼-inch-thick glass being 6mm.

Patterned glass is manufactured in a great variety of patterns, textures, and designs, all of which obscure vision and reduce light transmission. Among the popular patterns are ribbed, frosted, hammered, and fluted; it is also available tempered.

Tempered glass is heat treated for increased resistance to impact stresses. It is three to five times stronger than plate glass, much more costly, and pulverizes into small granules when broken. Tempered glass is used in doors, store fronts, and other areas accessible to the public. It may not be cut, drilled, or otherwise altered after manufacture.

Insulating glass (multi-glazing) consists of two or more sheets of glass separated by a hermetically sealed air space to provide thermal insulation, acoustical control, and an absence of condensation. Insulating glass can be manufactured from almost any type of glass, in exact sizes, and it may not be altered after it is manufactured. The seals may be glass-to-glass or metal-to-glass, and a reflective surface can be located on various faces of the insulating glass.

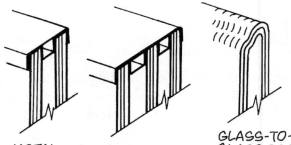

METAL-TO-GLASS SEAL   GLASS-TO-GLASS SEAL

INSULATING GLASS
Figure 6.15

Heat-absorbing glass (actinic glass) is tinted to absorb a high degree of solar radiation, and thus, it transmits less solar heat and glare into a building. It is usually bronze, gray, or blue-green in color.

Reflective glass consists of a reflective film laminated between two sheets of clear glass, or a thin film of metal oxide on the surface of the glass. It behaves somewhat like a mirror reflecting heat and glare, and thus, its performance is similar to that of heat-absorbing glass.

Laminated glass is composed of a plastic sheet sandwiched between two glass layers, all bonded under heat and pressure. When laminated glass breaks, the plastic sheet holds the broken glass in place, thus reducing the risk of injury in case of breakage. Laminated glass is also a better acoustic insulator than solid glass. There are two types: *safety glass*, which is used in automobiles and skylights; and *bulletproof glass*, which has four or more layers of glass laminated to three or more layers of plastic. This type is used where strong impacts may occur, as in banks and prisons.

Wired glass is made by placing wire mesh in the middle of glass during the manufacturing process. It has a high resistance to impact and remains intact after breakage. Wired glass is used where fire resistance is required and in skylights or other overhead glazing. It is available with a variety of different wire patterns.

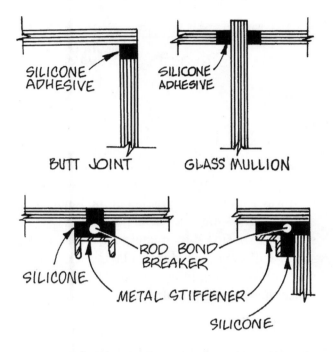

GLASS-TO-GLASS JOINTS
Figure 6.16

There are several modern glazing systems that render the metal framing members less visible. *Butt-joint glazing* consist of glass sheets with conventional head and sill frames, but no vertical mullions. A clear silicone sealant forms the vertical joint between adjacent glass sheets. In another system, the mullions are placed completely inside the glass, so that the windows appear flush and mullionless from the exterior. The glass is joined to the mullions with silicone sealant.

In addition to the glass types described above, there are several others that bear mention. *Mirrored glass* is glass with a metallic reflec-tive coating applied to one surface. *Psychiatric glass* (one-way glass) is seen as a mirror on one side and is transparent on the other side, as long as the light intensity on the mirrored side is greater than that on the other side. *Structural glass* is opaque and used for exterior building panels. *Corrugated glass* is used for decorative partitions, or in its wired form, for skylights and roof panels. *Glass block* is discussed in Lesson Two.

GLAZING PLASTICS

The term *plastic* refers to a vast range of synthetic materials. In this lesson, we are concerned only with those plastics that have, in recent years, replaced glass in applications such as skylights, translucent roof areas, fixed windows, interior partitions, luminous ceilings, and so on. For these purposes, acrylics, polycarbonates, polyesters, and polystyrenes are used. All of these materials are available in transparent, translucent, and opaque sheets, and all are either fire-resistant or classified as slow-burning.

Plastic, in flat sheets or corrugated form, can generally replace glass in most windows, skylights, shower enclosures, roof panels, or interior screens, where various transparent, obscure, or colored panels may be desired. However, plastic is not as hard as glass, it scratches far more easily, and thus, it is not as durable when exposed to normal use or weather. Plastic has a high coefficient of thermal expansion, and so the details must allow for a substantial amount of expansion and contraction due to temperature change. Plastic is normally more expensive than glass, but more resistant to fire and breakage. Another advantage of plastic is that it can easily be cut, bent, and formed.

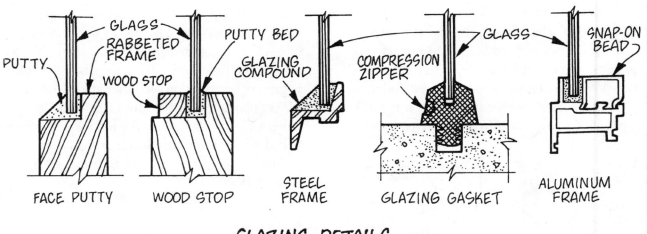

GLAZING DETAILS

Figure 6.17

GLAZING

Glazing is the process of placing glass or glazing plastics in windows and making a weather tight joint between the glass and its frame using various materials. Glass panels are set in the grooves designed to receive them, held in place with glazing beads, points, or clips, and sealed in place with various glazing compounds.

The various types of glazing compounds in general use are putties, elastic compounds, polybutene tape, polysulfide elastomer compounds, and structural glazing gaskets (compression or zipper types), which are molded or extruded shapes made from rubber, neoprene, vinyl, or other synthetics. These synthetics are particularly effective in allowing for the thermal expansion and contraction of glass.

Clearance must always be allowed between glass and frame so that the glass is surrounded by a watertight seal of glazing compound. In addition, glass must never be placed in tension, by exposing it to greater forces than it can safely withstand, nor should glazing take place when moisture might be trapped, which could cause the watertight seal to fail.

END DAY 3

LESSON 6 QUIZ

1. Which of the following glass types would one select in order to reduce light transmission?

   A. Insulating

   B. Patterned

   C. Reflective

   D. Laminated

2. Which of the following glass types are appropriate for skylight glazing?

   I. Laminated

   II. Tempered

   III. Wired

   IV. Heavy sheet

   A. I and III   C. I, II, and III

   B. II and III   D. I, II, and IV

3. Tactile finishes are generally applied to operating hardware in order to provide

   A. improved corrosion resistance.

   B. a more decorative finish.

   C. greater handling control.

   D. warnings to the handicapped.

4. The principal disadvantage of aluminum door and window sections is their

   A. high initial expense.

   B. lack of durability.

   C. poor resistance to galvanic action.

   D. poor resistance to building stresses.

5. What is the maximum height of a four-inch-wide wire-glass glazing strip located in a Class B labeled fire door?

   A. 13.5 inches

   B. 25 inches

   C. Full height of a 10-foot-high door

   D. No glazing is permitted

6. Which of the following are advantages of revolving doors?

   I. Emergency exiting

   II. Handicapped access

   III. Temperature control

   IV. One-way traffic flow

   V. Continuous pedestrian flow

   A. I and V   C. I, II, and IV

   B. III and V   D. III, IV, and V

7. A latching device that is completely concealed within a rectangular cavity carved in the edge of a door stile is commonly known as a

   A. mortise lock.   C. rim lock.

   B. unit lock.   D. cylinder lock.

8. Compared to glass, plastic sheet substitutes are generally less

   A. durable.   C. transparent.

   B. fire-resistant.   D. expensive.

9. Emergency exits in buildings often require the use of

  A. knurled knobs.

  B. lever handles.

  C. panic hardware.

  D. fusible links.

10. Which of the following qualities generally apply to wood windows?

  I.   Low maintenance

  II.   Low initial expense

  III. High resistance to fire

  IV.   High durability

  V.   Wide availability

  A. I and III   **C.** II, III, and V

  B. I, II, and IV   **D.** II, IV, and V

FINISH WORK

INTRODUCTION

Finish work refers to the methods, materials, and treatments that comprise the interior and exterior finished surfaces of a building. Finishes are all visible as an integral element of the completed structure; in fact, most of what one sees in a completed building is finish work that is superimposed over the building's structural framework. Interior finishes include floor finishes, such as paving, planking, and carpeting; wall finishes, including paneling, veneering, and tiling; and ceiling finishes, such as plastering, painting, and acoustical treatment. Exterior finishes consist of the entire outside surface of a building, including wall coverings, ornamentation, and protective coatings, such as paint.

The selection of a finish material is determined by its characteristics, cost, and aesthetic effect. Exterior surfaces must be weather resistant, durable, relatively free of maintenance, and architecturally appropriate in scale, texture, pattern, color, and so on. Interior surfaces must also be durable, easy to maintain, functional (considering fire, thermal, and acoustical qualities where appropriate), and aesthetically suitable. As most of what is visible in a finished building is finish work, it is obvious that architects must give serious consideration to the design, detailing, and execution of a structure's finish work.

Finish work has always been the element that has differentiated the refined from the crude. For example, in 17th century New England, the original settlers constructed primitive shelters, which were nevertheless successful in keeping the colonists relatively warm and dry through the first few harsh winters. Within a short time, having solved the basic functional problems, the early Americans demanded greater comfort and elegance in their shelters. Packed dirt floors gave way to wood planks, rough walls were paneled with finished wood boards, and eventually, building exteriors were covered with smooth boards or masonry, embellished with decorative details, and painted. It was this finish work that gave to early American architecture the distinctive look that has been so appealing for more than 200 years.

Many exterior materials were discussed in previous lessons; some of these will be reconsidered in this lesson as finish materials.

PLASTER WORK

Plaster is one of the oldest finish materials still used in construction today; more than 4,000 years ago it was employed by the ancient Egyptians. Greek and Roman builders also used plaster extensively; in fact, our word *plaster* comes from ancient Greece, where it originally meant "to mold." Even with the introduction of countless new materials, plaster remains one of the most popular surface finishes in almost every part of the world.

Plaster is a cementitious material that is applied in layers on both interior and exterior surfaces. It is composed of portland cement (for exterior plaster or stucco), or gypsum and lime (for interior plaster); an aggregate such as sand, vermiculite, or perlite; and enough water to form a workable paste. In some cases, hair, fibers, or mineral aggregates are added to impart some desirable quality. Plaster mixtures are applied in two or three coats over a base of concrete or masonry, metal lath, or various types of lathing board.

In addition to portland cement plaster, which is used on the exterior, and gypsum plaster, which is used on the interior, several other types of plaster are available as follows:

Acoustic plaster—provides acoustical treatment for walls and ceilings

Bonding plaster—used on interior concrete walls and ceilings

Fire-resistant plaster—provides fire resistance for steel and other materials

Keene's cement plaster—contains lime putty for hard, water-resistant finish

Lightweight plaster—contains vermiculite or perlite aggregates for fire resistance

Bases for plaster include concrete, masonry, and lath. Concrete and masonry surfaces must be sufficiently rough and porous to provide a good bond; in lieu of such a base, lath should be spread over the surface, on furring channels, prior to plastering.

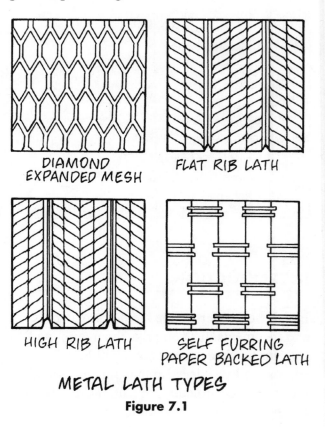

DIAMOND EXPANDED MESH   FLAT RIB LATH

HIGH RIB LATH   SELF FURRING PAPER BACKED LATH

METAL LATH TYPES

Figure 7.1

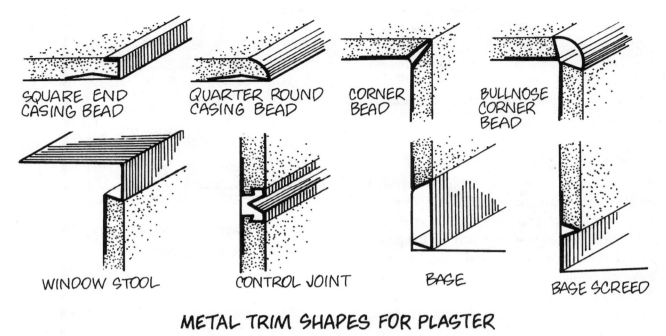

SQUARE END CASING BEAD QUARTER ROUND CASING BEAD CORNER BEAD BULLNOSE CORNER BEAD

WINDOW STOOL CONTROL JOINT BASE BASE SCREED

METAL TRIM SHAPES FOR PLASTER

Figure 7.2

Metal lath is a sheet metal or wire fabric into which a base coat of plaster is keyed. It is fabricated from copper alloy steel coated with a rust-inhibiting paint or from galvanized steel, and it is specified in pounds per square yard. Metal lath is available in several types, the most common of which are flat expanded (diamond mesh), rib lath (flat rib or high rib), and woven wire mesh or lath. When plastered, the metal lath is completely embedded, which creates a mechanical bond between plaster and metal.

Lathing board, sometimes referred to as plasterboard, gypsum lath, or gyplath, is composed of gypsum that has been mixed with water, hardened and dried, and sandwiched between two sheets of heavy, porous paper, which provides adhesion for the plaster. The standard size of lathing board is 16" × 48", and it is either ⅜ or ½ inch thick. Lathing board is available in solid sheets, perforated sheets (for better mechanical keying), in solid sheets with reflective aluminum foil on one side (for vapor and thermal control), and in Type X, which provides a fire-resistive rating.

Lathing board is generally nailed in a horizontal position to wall framing members that are spaced 16 inches on center. Vertical joints should be staggered and arranged to coincide with studs, blocking, and so forth, to which the board is solidly nailed.

Before plaster is applied, a variety of trim shapes are set, such as metal corner beads and casing beads for the protection of corners and edges, and plaster grounds and screeds, which ensure level and plumb surfaces, as well as a consistent thickness throughout. Interior plaster over metal lath should be a minimum of ⅝ inch thick, and over gypsum lath, ½ inch thick. Exterior plaster varies from ¾ to ⅞ inch thick, depending on the base.

Plaster is generally mixed mechanically and applied in three coats, referred to as *scratch, brown,* and *finish* coats. In two-coat work, the scratch and brown coats are combined in a base coat. The proportions of a conventional gypsum plaster, applied over metal lath, are shown in Table 7.1.

Coat	Sand/Plaster	Thickness
1st - Scratch	2 parts to 1	¼ inch
2nd - Brown	3 parts to 1	¼ inch
3rd - Finish	1½ parts to 1	⅛ inch

Table 7.1

Plaster is applied by hand in a variety of textures, or it may be machine applied to create a mechanically textured finish. Plaster must be cured, very much like newly poured concrete. It should never be applied when the temperature falls below 40°F, and in the summer, rapid evaporation must be avoided.

Veneer plaster is a specially formulated dense plaster that is applied in a thin coat over gypsum lath. It may have a smooth or textured finish and can usually be painted within 24 hours after plastering.

GYPSUM BOARD

Gypsum board is a prefabricated form of plaster used in place of conventional three-coat interior plaster. Because it is usually applied in large, single-thickness sheets, its application is much faster than using several coats of wet plaster. In addition, it is generally cleaner to use, requiring less protection of other work and less final cleanup.

Gypsum board consists of a gypsum plaster core reinforced with paper or other materials laminated to both sides. It is commonly referred to by the popular trade name *sheetrock*, or by the term *drywall*, because—unlike plaster—it is applied in dry sheets. Gypsum board is manufactured in 4-foot widths, from 6 to 12 feet in length, and from ¼ to ⅝ inch in thickness (thin sheets may be used on curved surfaces).

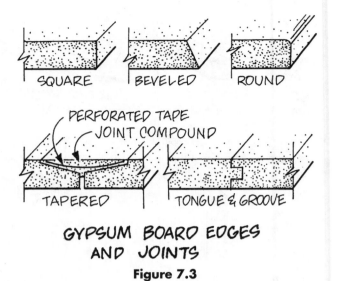

GYPSUM BOARD EDGES AND JOINTS
Figure 7.3

Gypsum board is applied directly to wood or metal framework or to concrete or masonry walls above grade with nails, screws, or adhesives. Exposed corners and edges are protected with metal trim, and all joints, nailhead depressions, and other dents are covered with a perforated tape embedded in joint compound. When the compound has hardened, the surface is sanded smooth and painted or overlaid with a wall covering. A prime coat is often applied to gypsum board in order to prevent raising a nap on the paper surface.

The following is a list of some common types of gypsum board with a description of their uses:

Regular board. Surfacing for walls and ceilings.

Backing board. Backing for other materials, such as acoustical tile.

Insulating board. Aluminum foil laminated to one side, which serves as reflective insulation and vapor barrier.

Type X board. For fire-resistive rating.

Moisture-resistant board. Backing for ceramic tile in high moisture areas.

Decorative board. Prefinished with decorative paper or vinyl in various textures, patterns, and colors.

CERAMIC TILE WORK

Ceramic tiles are small, flat units composed of clay or clay mixtures fired in kilns at high temperatures. They are set on floor, wall, or ceiling areas to provide permanent, waterproof, durable, and easily maintained finish surfaces. Ceramic tile may be installed on the interior or exterior of a building, depending on the density of the tile, its finish, and its method of installation. Ceramic tile is commonly available in square, rectangular, hexagonal, and numerous other shapes; and its size varies from about 1 to 9 inches in length and width, and from ¼ to 1 inch in thickness. Trim tiles are available in various shapes for use as corners, bases, coves, caps, and other such mouldings.

Ceramic tiles are manufactured by burning clays that have been pressed into the desired shape through either the dust-pressed process or the plastic process. Vitrification, or the fusion of the tile after firing, is a measure of tile density, and the common classifications are listed below:

Non-vitreous tile has a moisture absorption rate between 7 and 15 percent.

Semi-vitreous tile has a moisture absorption rate between 3 and 7 percent.

Vitreous tile absorbs less than 3 percent of its weight in moisture and has a density such that it cannot be penetrated by dirt.

Impervious tile repels almost all moisture and dirt. It is extremely hard and generally available only by special order.

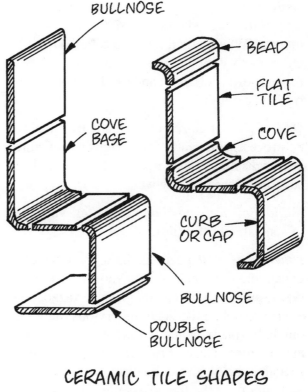

CERAMIC TILE SHAPES
Figure 7.4

Finishes for ceramic tiles vary greatly, but essentially, they are either glazed or unglazed. Glazed tiles have a glassy, waterproof surface produced by ceramic material that is fused on their faces. The glazes may have several different textures, or degrees of matte finish, and they normally range from pure white to jet black and every hue in between. Unglazed tiles derive their color and texture from the clays used in their manufacture, or from minerals that are added prior to firing.

Some common types of ceramic tile are listed below:

Glazed interior tile. Non-vitreous, made by the dust-pressed process and used for interior surface applications.

Glazed weatherproof tile. Semi-vitreous or vitreous, may be used for moderate wear on the exterior.

Ceramic mosaic tile. Generally unglazed, semi-vitreous, and less than six square inches in area. (When larger, they are called pavers.) Often factory-mounted on large paper sheets for ease of installation, and used on both interior and exterior.

Quarry tile. Unglazed units, made by the plastic process, and nearly impervious to moisture, dirt, and stains. Weatherproof, durable, and suitable for heavy duty wear.

Inlaid, Faience, and Handmade tiles. Varieties of special, colorful tiles.

Other types of tile. The tile industry is rapidly expanding to include a wide variety of tile types, including porcelain, glass, terrazzo, and stone. Porcelain tile is an impervious tile without a clay backing and is suitable for exterior applications. The color is throughout the tile although the glaze can be different on the top and bottom of the tile. Glass tile is made from glass that can be "tumbled" to ease the edges and sometimes has color coatings, such as gold foil, on the back. Terrazzo tile is a manufactured tile consisting of an aggregate and cement matrix similar to terrazzo. Lastly, stone less than ¾-inch thick is generally referred to as stone tile and is usually installed in the same manner as ceramic tile, including thin bed methods not typical for dimensional stone.

Setting tile requires great care; backing surfaces must be rigid, water-resistant, and level or plumb. Poorly prepared surfaces or any subsequent structural movement can lead to cracking at the grouted joints and possible leakage. Ceramic tile is frequently set in a cement mortar bed about one inch thick and composed of one part portland cement, half part hydrated lime, and five parts sand. Tile may also be set with the use of adhesives (thinset), such as latex-portland cement mortar, epoxy mortar, or a mastic called organic mortar.

After tiles are cemented in place, they are grouted and pointed (to shape and compress the joint) with a rich mixture of cement and hydrated lime. The grout is worked into the joints, which range from $\frac{1}{16}$ to $\frac{1}{2}$ inch wide, and the entire surface is wiped clean after completion.

Ceramic tile is also available in prefabricated panels that fit standard tub and shower base dimensions. These prefabricated units use a plastic adhesive as the joint filler.

PLASTICS

In recent years, a great many plastic products have replaced building elements formerly made from other materials, such as flashing, flooring, countertops, etc. One such substitute material is *plastic siding.* It is made from a vinyl plastic and is available in various colors, textures, with or without insulation backing. Other items include *plastic countertops* with a molded-in lavatory, in a wide range of colors and marbleized designs; *plastic bath tubs* that include full-height enclosures; and *plastic shower stalls* consisting of integral base and walls. All of these are manufactured from various acrylics or fiberglass.

FLOORING

Finish flooring is the surface on which people or equipment move within a building. It covers a vast range of treatments, and it includes virtually every material that can be walked on, from a painted concrete slab to a custom-designed wood parquet floor.

Finish flooring in a building involves more than just covering up the rough construction. Materials such as ceramic tile, terrazzo, and concrete produce highly durable finishes that are capable of withstanding constant, heavy traffic. Cork, rubber, and carpet control the noise level in a room, vinyl is grease-resistant, brick is fire-resistant, and synthetic resin surfaces require very little maintenance. It should be apparent, therefore, that the selection of a finish flooring material is not a haphazard choice; rather it is based on occupancy, location, anticipated use, initial cost, resistance to fire, noise, and dirt, frequency and cost of maintenance, comfort, and finally, appearance and personal preference. Several common flooring materials are discussed below.

Wood Flooring

Wood flooring is a popular finish suitable for a wide range of uses. It is durable, comfortable, attractive, and easily maintained. Wood flooring is manufactured under strict rules that control the grade and moisture content of the wood. Close-grained, hard, and durable species of both hardwood (oak, maple, birch, beech, pecan) and softwood (pine, fir, spruce, hemlock, redwood) are used for flooring. Flooring may be plain-sawed or quarter-sawed, but vertical grain boards are preferred for the best installations. The basic types of wood flooring are strip, plank, and block, which are usually fastened to the subfloor by means of nails, screws, or mastic cements; all are available prefinished or unfinished.

All wood is susceptible to swelling and shrinkage caused by changes in humidity, and therefore, sufficient expansion space must be provided at a floor's perimeter. In addition, wood flooring used in humid climates should be treated against decay, mold, and insects.

Strip flooring is available in various thicknesses ranging from $5/16$ to $25/32$ inch, in widths ranging from $1\frac{1}{2}$ to $3\frac{1}{2}$ inches, and in random lengths.

The strips most often have tongue-and-groove sides and ends. Strip flooring is applied over a wood subfloor or on spaced wood sleepers (horizontal members that elevate the finish floor above the subfloor below), and is generally fastened by blind nailing. It may also be applied directly to a concrete slab using waterproof adhesives. Strips are laid in conventional parallel lines, in various simple patterns, or in intricate parquet designs.

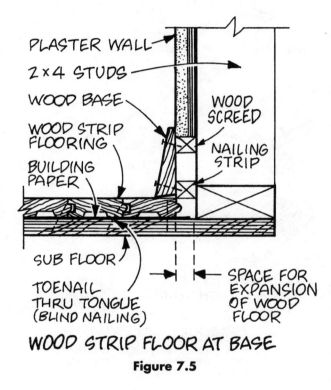

WOOD STRIP FLOOR AT BASE

Figure 7.5

Plank flooring is available in the same thicknesses as strip flooring, but in widths between $3\frac{1}{4}$ and 8 inches. It is frequently laid in random widths with a V-groove where the board edges meet. Wide planks are susceptible to warpage, and therefore, they are often screwed to the subfloor through the face of the plank, which is then plugged and finished smooth. The wood plugs result in a distinctive appearance that has become the distinguishing mark of wood plank floors.

Parquet flooring consists of prefabricated squares measuring about 9 to 12 inches square and between ½ and 25/32 inch thick. They may be either solid strips held together with splines or plywood veneers, and they are available unfinished or prefinished, with square or tongue-and-groove edges. Thin block flooring is fastened by nailing or by adhesives.

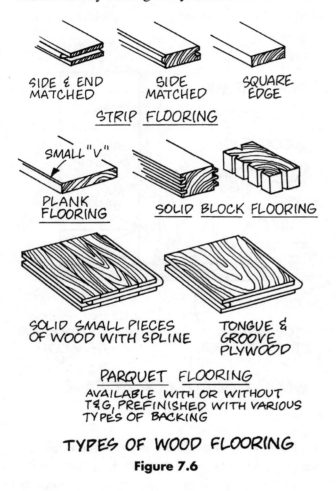

TYPES OF WOOD FLOORING

Figure 7.6

Solid block flooring consists of end-grained blocks that have been treated with creosote against moisture and decay. The blocks are usually set in a coating of bituminous material, and they are used in workshops and industrial buildings where a durable, heavy-duty floor is required. This type of flooring is very resistant to oil, grease, and mild chemicals.

Mortar-Set Flooring

Mortar-set flooring includes any masonry material that is set and utilized as a finish floor surface. Examples are tile, brick, slate, and other types of natural stone.

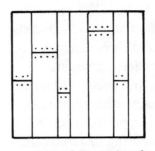

WOOD FLOORING PATTERNS

Figure 7.7

Brick of almost any type can be used as finish flooring, depending on the desired result. Acid-resistant brick is dense, hard, and often used for industrial flooring. Paving brick is also dense and hard, and it provides an attractive finish in a variety of natural colors. Building brick is warm and attractive, in a rustic way, but it is generally too porous and uneven for most interior floors. Brick is normally laid in a mortar setting bed over the subfloor; however, it may be laid with thinset mortar or adhesives. On the exterior, brick is often set in sand with tight joints.

Slate is very durable, attractive, and expensive. It may be slightly rough or relatively smooth, rigidly rectangular or completely irregular, and it is available in a rich range of natural colors. Slate paving is about one inch thick and generally laid in a mortar setting bed, in which case the joints are grouted and finished flush. However, it can also be laid with adhesives or set in sand on the exterior.

Stone flooring, other than slate, includes limestone and sandstone (flagstone) as well as granite and marble. All stone used for flooring is about one inch thick and has a relatively smooth surface. Stone is hard, durable, and heavy, and regardless of type, it is easily stained.

Poured-in-Place Flooring

In addition to concrete, which can be finished with stain or paint, poured-in-place flooring includes terrazzo, magnesite, and a variety of synthetic seamless floors.

Terrazzo is a concrete topping consisting of marble chips in a cement or synthetic resin binder, ground and polished to a smooth finish. Terrazzo is dense, permanent, hard-wearing, and relatively expensive. The topping is about ⅝ inch thick and generally laid over a 2½-inch-thick concrete setting bed; this, in turn, may be applied over rough concrete, wood, or steel decking.

A disadvantage to terrazzo flooring is its tendency to crack, but this may be alleviated by the use of regularly-spaced divider strips. These are available in plastic or a variety of metals that are partially embedded in the concrete setting bed. The crushed marble mixture is then poured between the divider strips, compacted, troweled smooth, and machine-ground to a smooth finish. When using a synthetic resin binder, dividers are unnecessary and often omitted.

Terrazzo floors are available in a wide variety of colors, depending on the color of the cement and marble chips used. It may also be produced with an abrasive finish, as a conductive floor (for hospitals), and in several prefabricated shapes, such as tiles, stair treads, and coved bases.

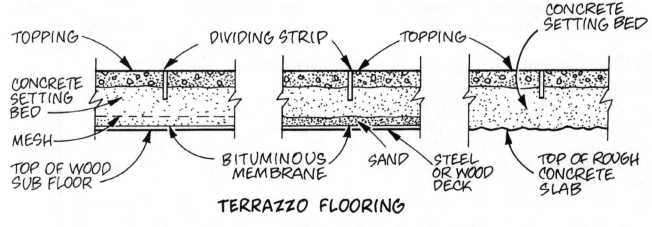

TERRAZZO FLOORING

Figure 7.8

Magnesite seamless floors (magnesium oxychloride) are made from a composition of magnesium oxide, sand, and magnesium chloride, which forms a plastic material that is troweled to a smooth finish. The finished surface, ranging in thickness from ½ to 1 inch, may be applied directly over wood, concrete, or flat steel decking. Magnesite composition flooring is resistant to stains and fire, somewhat resilient, and relatively inexpensive. It is appropriate for use in schools, offices, and industrial buildings.

Elastomeric floors are composed of clear urethane resins applied as a liquid over an elastomer underlayment. They are generally hard, durable, and relatively easy to maintain. Colored vinyl chips are often embedded in the urethane for their decorative effect.

Rubberized plastic floors consist of several layers of various neoprene compositions troweled over a waterproof latex membrane. The finish is smooth, durable, and completely waterproof. This type of flooring may be used on the interior or exterior of a building.

Paint-type floor finishes consist of a synthetic resin (epoxy, urethane, polyester, etc.) and an activator. These, when mixed, are applied directly to concrete, wood, or metal to produce an exceptionally hard, durable, wear-resistant, easily-maintained, greaseproof floor finish.

Resilient Flooring

Resilient flooring materials are durable, non-absorbent, comfortable, and easily maintained. They are thin (¹⁄₁₆ to ³⁄₁₆ inch thick) and available in tiles (generally 9 or 12 inches square) or in six-foot-wide rolls. These materials are laid over smooth wood or concrete subfloors and cemented in place with mastic cements. Trim shapes such as topset wall bases, stair nosings, and edging strips are generally available. Some common types of resilient flooring material are described in the following paragraphs.

Vinyl is available in sheet or tiles and may be used above or below grade. It is very resilient; durable; resistant to indentations, oil, and water; and easy to maintain. Therefore, it is particularly appropriate for kitchens. Available in a large assortment of colors and patterns, it is moderately expensive.

Asphalt tile is relatively inexpensive and very durable, but it lacks resilience. Asphalt tiles are easily installed above or below grade in areas of heavy traffic, such as in schools, hospitals, and offices.

Linoleum is durable, resilient, easily maintained, and relatively inexpensive, considering its long life. It is available in printed patterns, inlaid designs, or solid colors throughout its thickness. However, it is unsuitable below grade or in the presence of subsurface moisture.

Rubber flooring is now made of synthetic rubber. It is durable, elastic, quiet underfoot, easily cleaned, and moderately expensive. While it is attractive and comfortable, it cannot resist oils or greases.

Cork flooring is available in sheet or tiles with a synthetic resin binder and surface. It is highly resilient, resistant to staining, durable, and easily maintained. It also has excellent acoustical properties.

Carpeting

Carpeting, once used almost exclusively as area rugs in homes, now includes carpet tiles and wall-to-wall carpeting, which are commonly installed in many commercial and institutional buildings. They add warmth, resilience, texture, and effective control of noise. Carpeting

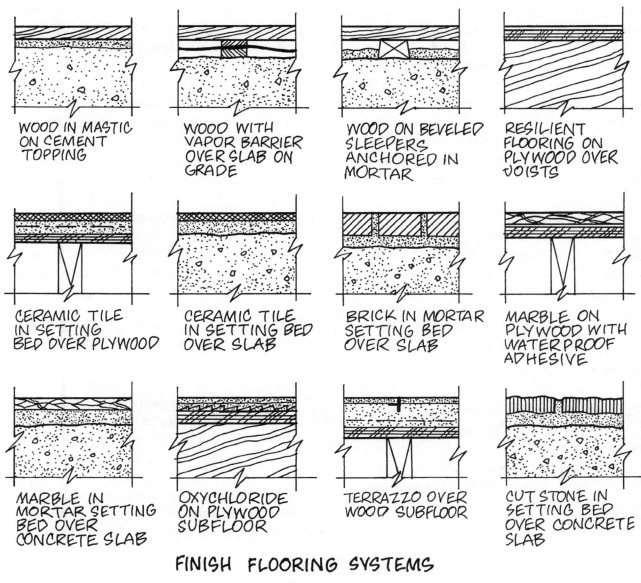

FINISH FLOORING SYSTEMS

Figure 7.9

eliminates the need for a finish flooring material, as it can be installed directly over a plywood subfloor or a concrete slab. It may also be used on wall surfaces to improve the acoustical quality of a space, and also to prevent accidents in institutional buildings, such as nursing homes.

Carpeting is classified by material (acrylic fibers, nylon, polyester, wool), by method of manufacture (tufting, weaving, needle punch-

ing), and by texture (low pile, high pile). Most commercial grades of carpeting are durable, soil-resistant, and relatively inexpensive to maintain; however, their ultimate durability depends on several factors. With pile carpets, density of pile (tufts per square inch) is of primary importance; the longer and denser the pile, the longer the life of the carpet. Cushions or padding installed beneath the floor covering will also add years of life to carpeting. However, both carpet tiles and wall-to-wall carpeting are available with an integral foam backing, which eliminates the necessity for padding.

All carpeting requires maintenance, although, in general, it needs less care than resilient flooring. Embedded dirt can harm carpeting and it should therefore be shampooed periodically. Depending on the fibers and treatments used in their manufacture, carpets can be made mothproof, mildew-resistant, flame-resistant, stain-resistant, and antistatic.

The initial cost of carpeting is relatively high. However, due to its moderate cost of maintenance, it is often the most economical floor covering to use, when considered over its entire life cycle.

ACOUSTICAL WORK

Acoustics, the science of sound and its control, is beyond the scope of this lesson; however, as materials and systems used for sound control are increasingly used for interior finishes as well, we shall consider their use as another aspect of finish work.

A wide range of acoustical materials are used to control or minimize sounds. The criteria in their choice include not only acoustical efficiency, but also fire resistance, method of application, ease of maintenance, and appearance. Almost all acoustical materials control sound by absorption, and they may be classified as follows:

Acoustical tiles are manufactured from wood, mineral, or glass fibers with perforated, textured, or patterned surfaces that allow sound to penetrate. They are lightweight, fragile, and most often factory finished. Available in one, two, and four-foot modules, they have square, beveled, rabbeted, or tongue-and-groove edges.

Acoustical panels are made of perforated steel, aluminum, fiberboard, or hardboard, which is backed with a soft, sound-absorbing batt or blanket. They are permanent, far more durable than tiles, and have acoustical properties that depend on the quantity and size of the perforations and the sound absorption value of the pads used behind the panels.

Prefabricated tiles and panels are often nailed or cemented to a solid backing surface. Because tiles are so fragile, they are almost

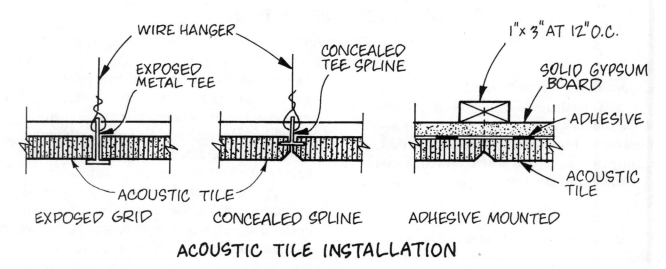

ACOUSTIC TILE INSTALLATION
Figure 7.10

always used for ceilings, and frequently, acoustical tile ceiling systems are suspended from the structure above. This creates a plenum for mechanical conduits and recessed lighting fixtures. When a fire-rated ceiling is required, acoustical tile supporting systems must have an approval from the Underwriters' Laboratories.

Wet materials include acoustical plaster and mineral-fiber products to which a binding agent is added. These may be hand troweled or machine sprayed to produce an incombustible surface, usually with both acoustical and thermal insulation values. Especially suited for irregular or curved surfaces, these materials are quickly installed and very effective.

Vibration control consists of isolating or reducing mechanical vibrations in or near a building. To control vibration, dense materials such as lead, solid concrete block, or concrete must be used to isolate the vibration source from the building spaces.

Several methods are used to reduce sound transmission through walls. These include staggering wall studs or using two separate sets of studs, mounting the finish material on resilient furring channels, and using fiberglass or other acoustical insulation.

PAINTING

The use of paint goes back to prehistoric times. It was used some 25,000 years ago in the famous cave paintings at Lascaux, and perhaps even earlier to decorate earthenware pots. Throughout prehistory, paint served to ornament various elements used for ceremonial or religious rites, and until comparatively recently, in fact, paint was used almost exclusively to embellish or beautify buildings.

Today, the principal purpose of paint is to protect and preserve the material being painted and secondarily, for aesthetics. Originally, paint consisted of a simple mixture of white lead and linseed oil that was applied in three to six coats over a long period of time. Modern technology, however, has changed all that; some synthetics are capable of covering in one coat and drying in only a half-hour. Additionally, because of the danger of lead poisoning, titanium dioxide and other "safe" pigments have completely replaced white lead.

An incredibly vast range of materials is now available to treat finish surfaces; paint can resist the damaging action of weather, moisture, mildew and decay, it can make surfaces more sanitary, it can modify the light in a space, and it can produce a myriad of psychological effects through the use of color and texture. Paint, incidentally, is only one of many organic coatings, which is the proper term applied to all types of protective liquid coverings.

Paint is a protective coating that combines a pigment and a vehicle. When applied to a prepared surface, this mixture forms a solid film that acts as a barrier between the material painted and those elements that may adversely affect it.

Pigments are finely ground solids that are held in suspension by the vehicle. They provide the paint's color, as well as its *hiding power,* or opacity. Hiding power is a function of the difference in indices of refraction of the pigment and the vehicle. A large difference produces good hiding power and also minimizes the damaging effect of sunlight, as much of it is reflected. Colored pigments are finely ground solid particles made up largely from inorganic materials, such as iron oxide that produces yellows and sodium silicate that produces blues. Synthetic color pigments are also available that widen the range to include the total color spectrum. Colored pigments differ from white pigments in that light is selectively absorbed. Extender pigments are additives used in a paint mixture to control gloss, increase durability, improve workability, and reduce cost.

Vehicles, which constitute the liquid portion of a paint mixture, consist of a binder that forms the paint film, driers to speed up formation of the film, and solvents or thinners that control consistency and aid in drying. Depending on the type of solvent used, a paint will form a solid film by oxidation, evaporation, or by thermosetting action caused by applied heat. Vehicles contain drying oils or resins, alone or in combination. Driers, such as linseed oil, have the ability to absorb oxygen and change into a solid state. Resins that are natural (varnish) or synthetic (phenolics, epoxides, alkyds, latexes, urethanes, polyvinyls) all add to a paint's durability and protective quality.

Paint types are numerous and widely varied. They may be classified by location (exterior enamel); by the material on which they are applied (wood stain); by their finish (semigloss enamel); or by their unique characteristics (lacquer). Following is a brief review of several types of coatings.

Pigmented Coatings

Pigmented coatings consist of white pigment, to which color may be added, and a vehicle. They include house paints, metallic paints, colored lacquers, water-based paints, and synthetic resin paints. These are generally available ready-mixed for immediate use.

- *Enamels* are pigmented paints that use varnish as the vehicle. They form a hard, tough film that is durable and resistant to weathering. Enamels are available in a gloss or matte finish.

- *Baked enamels* are always factory-applied, as they require a controlled temperature of between 200 and 300°F to harden. A baked enamel finish is hard, durable, washable, and resistant to most mild chemicals.

Clear Coatings

Clear coatings protect and beautify surfaces without obscuring their natural appearance. Although preferences still vary, environmental concerns have resulted in major changes as the coating industry attempts to reduce volatile organic compound (VOC) emissions.

Varnishes using volatile solvents such as turpentine and mineral spirits have been replaced by polyurethanes and acrylic modified urethane coatings. Lacquers made from a highly flammable nitrocellulose base have been replaced by acrylic lacquers. And shellac, an alcohol-soluble resin formerly used as a sealer or prime coat, has been replaced by water-based primer-sealers that are more durable and less susceptible to water spots. Shellac as a floor finish has been replaced by acrylic modified urethanes.

Clear sealers include a variety of systems that prepare surfaces for another finish. Water-borne sealers and water-repellant compounds are now readily available using acrylic, urethane, and other binders, often with better results than the less environmentally gentle sealers they replace.

Stains, although not strictly clear coatings, have a pigment content that is lower than opaque paints. They do not obscure the natural grain of wood or substrate. They also have low viscosity and deep penetration; because the stain is absorbed by the substrate rather than forming a film on its surface, the substrate and its grain remain visible.

Bituminous Coatings

Bituminous coatings are made from coal tar and asphalt, and are used as a protective coating for submerged ferrous metal and for waterproofing masonry materials.

- *Coal tar pitch* must be melted to a fluid for application. It is used to protect metal and is applied by dipping. Coal tar mixtures may be applied hot or cold and are used as waterproofing agents.

- *Asphalt* coatings are available as paints, enamels, and emulsions. In comparison to coal tar products, they are more weather resistant and less affected by temperature extremes. They are commonly used for dampproofing and waterproofing work.

Miscellaneous Coatings

Beyond the coatings discussed elsewhere in this lesson, numerous products exist for more specific applications. Care should be taken to establish the suitability of the coating for both the substrate and the application.

- *Cement* coatings consist of Portland cement and, occasionally, lime (with water as the vehicle). They may also contain pigments, water repellents, or fillers. Cement coatings are used for damp-proofing masonry and concrete surfaces, and for steel protection.

- *Fire-resistant* and flameproof paints (intumescent paints) do not support combustion and have developed to the point where they have fire ratings that can be used to protect structural steel in fire-rated construction. They can also be used to improve flame-spread ratings and stop heat transfer to combustible materials.

- *Rust-preventive paints and primers* are characterized by their rust-inhibitive qualities, low permeability to corrosive elements, and low water absorption. They are available as primers and as finish paints, and typically include zinc or (in some cases) portland cement as the principal ingredient.

- *Antibacterial* and insecticidal paints are produced by adding counteracting ingredients to the paint. Some ingredients lose effectiveness over time and some may have toxic effects harmful to plants and animals.

- *Epoxy coatings* resist chemicals, moisture, or stains. Both one-part and two-part coatings are available.

- *Specialty coatings* are being promoted by paint manufacturers for a wide variety of purposes and can provide the design team with options not previously available. Care should be exercised, however, to verify manufacturers' claims, preferably by visiting local applications of the product to judge its success. Where no local application is available, another option might be a better choice.

Paint Application

Paint application requires knowledge, skill, and understanding of painting materials and methods. Paint is applied with brushes, spray guns, rollers, trowels, or rags at the job site, or by dipping, tumbling, or spraying at the mill, shop, or factory. Frequently, materials receive a prime coat at the shop and are finish painted at the job site.

All materials receiving paint must be properly prepared to insure adhesion of the film coating and maximize the life of the coating. Surfaces must be clean, dry, and free of all corrosion, grease, or surface defects that will prevent adhesion. Paint should be applied at temperatures between 55 and 85°F, for best results.

In addition to selecting the paint, architects must specify the surface preparation (such as sealing knots in wood), the method of application (brush, rollers), and the number of coats desired (prime coat, finish coat, and so on).

All painted surfaces must eventually be maintained or refinished, but paint failures, such as blistering, scaling, and peeling, will be minimized if surfaces are free of all moisture, well prepared, and finished with the appropriate coating.

LESSON 7 QUIZ

1. Which of the following finish floorings would one choose if the most important criterion was resilience?

    A. Elastomeric-type flooring
    B. Asphalt tile flooring
    C. Cork sheet flooring
    D. Terrazzo with synthetic resin binder

2. Vermiculite is sometimes added to a plaster mix in order to enhance the mixture's

    A. durability.
    B. workability.
    C. water resistance.
    D. fire resistance.

3. Which of the following are advantages of using a suspended acoustical tile ceiling system?

    I. Light fixtures may be recessed.
    II. Mechanical conduits may run above the ceiling.
    III. Enclosure provides a one-hour fire rating.
    IV. Acoustical tiles provide a tough, durable ceiling.
    V. Almost all systems are conveniently modular.

    A. I and IV   C. II, III, and IV
    B. I, II, and V   D. I, II, III, and V

4. In a remodeling project, there are several rough concrete basement walls that require a smooth plastered finish. Before plastering, a plumb surface should be created using

    A. gypsum lath.
    B. gypsum board.
    C. plywood.
    D. metal furring channels.

5. All the following are advantages of using gypsum board in place of wet plaster *EXCEPT*

    A. it is generally less expensive.
    B. it is generally more durable.
    C. it is cleaner to use.
    D. it is faster to use.

6. Painting failures are *MOST* often caused by

    A. poor quality paint materials.
    B. faulty application methods.
    C. surfaces that are not clean and dry.
    D. painting when the weather is hot and humid.

7. Both Keene's cement plaster and ceramic tiles are suitable to use on the walls of an institutional shower room. One might select the plaster rather than tiles because it is

    A. more water-resistant.
    B. more durable.
    C. less expensive.
    D. less difficult to maintain.

8. Which of the following apply to the scratch coat of plaster?

    I. It is generally the first plaster coat.

    II. It consists of a mixture of portland cement, or gypsum and lime, with water and sand.

    III. It is generally applied over gypsum lath.

    IV. It is always applied by hand.

    V. It is rarely less than ¾ inch thick.

    A. I and II

    B. I, II, and III

    C. III, IV, and V

    D. I, II, III, and IV

9. The major disadvantage of wood flooring is its susceptibility to

    A. weather-related swelling and shrinking.

    B. indentations from heavy traffic.

    C. damage from fire.

    D. stains from oil, grease, and chemicals.

10. For which of the following reasons would a urethane coating be preferred?

    A. It dries quickly.

    B. It affords a high degree of protection.

    C. It prevents rust.

    D. It provides resistance to fire.

BUILDING CODES

by John A. Raeber, FAIA, FCSI, CCS

INTRODUCTION

One of the first steps in designing any facility is to determine governmental requirements applicable to the construction of the facility. Building construction is limited by a variety of regulatory conditions that are typically enumerated in various codes adopted by agencies charged with specific tasks relating to public health, safety, and welfare.

Building codes define the allowable size of a building, both in area and in height, based on the degree of danger defined by the type of occupancies or uses anticipated in the building and on the types of construction. A building structure constructed of combustible materials would be limited in both size and height. A building structure constructed of noncombustible materials with proper fire ratings and a fully automatic sprinkler system could be built as large as the budget allows.

In addition, building codes define requirements for fire ratings, fire protection, egress, and accessibility. They also provide specific requirements for materials and methods of construction, usually prescriptive in nature, but sometimes related to performance characteristics such as structural design requirements for dead loads, live loads, snow loads, wind loads, rain loads, and seismic loads.

Specific information provided here relating to building codes is intended to be conceptual in nature and should be relevant to most building codes. In many cases, identifiers are taken from the International Building Code but are consistent with recent editions of the original three model building codes used to create the IBC. The NFPA 5000 generally recognizes the same concepts defined here but may use different methods for identifying specific elements such as not assigning specific letters to occupancies (for example, Group A as Assembly, Group B as Business) and there are differences between the IBC and NFPA 5000 relating to specific requirements throughout the codes.

HISTORY

Ancient history presently suggests the Babylonian Code of Hammurabi, written some 4,000 years ago, as the original building code that stated if a building collapsed and killed the owner, the builder would be put to death. In modern times, starting in the 19th century, requirements limiting the type of construction and sizes of buildings were originally defined by insurance companies. Losses of both life and property in buildings caused building owners to seek insurance, and insurance companies like the American Insurance Association, developed requirements to minimize their losses. In the early 20th century, communities saw the benefits of these requirements and began using

them as the basis of construction requirements for their communities.

Over time, officials involved in developing and enforcing building codes joined together in associations and began developing regional building codes that reflected the needs of their areas. One group of building officials out of the Chicago area developed a model building code with an emphasis on snow loads. Another group of building officials out of the Atlanta area developed a model building code with an emphasis on wind loads, especially the extreme winds in hurricanes. Yet another group of building officials out of California developed a model building code with specific seismic load requirements.

As communications and travel improved, these groups shared their specific knowledge and many of the specific requirements relating to snow, wind, and seismic loads were identical in the three model building codes. Modifications were made to bring the three model building codes into a common code format and to minimize the differences. Still, requirements relating to the allowable size of the building and other requirements did vary between the model building codes until the adoption of the International Building Code in 2000 by the International Code Council.

Unfortunately, in 2002, the National Fire Protection Association (NFPA) published a series of codes referred to as the Comprehensive Consensus Codes (C3) including NFPA 5000, "Building Construction and Safety Code." So, architects are once again faced with the possibility of local communities' adopting either of the two model building codes, the IBC or the NFPA 5000. The IBC seems to be the preferred model building code at this time but architects must verify which building code has been adopted as well as which edition of the code.

The International Building Code presently has both an IBC 2000 Edition and an IBC 2003 Edition. The NFPA 5000, 2003 Edition is the first edition but will, like the IBC, be updated on a regular basis.

USE AND OCCUPANCY

The intended use or "occupancy" of a building provides the first major design consideration. Each occupancy relates to a specific type of concern such as the potential for panic in facilities where large numbers of people congregate, special concerns when there are young people in need of adult guidance in an emergency, concerns for the safety of people who might be sleeping, and facilities that might have special hazardous materials.

The International Building Code classifies ten major types of occupancies, with some having from two to five subclassifications. NFPA 5000 has similar occupancies but does not use the abbreviations (A, B, E, etc.).

Group A—Assembly: In facilities where more than 50 people gather for a common purpose such as entertainment, recreation, food, or drink, there is potential that irrational action by one person could result in irrational action by many. This irrational action is called panic and one of the requirements of the assembly occupancy is that doors are required to swing in the direction of exit travel and to have panic hardware allowing the pressure on the face of the door to release any latch device.

Group B—Business: Professional and service facilities as well as drinking and dining facilities with fewer than 50 occupants are the safest space available because most of the occupants are familiar with their surroundings and are expected to be awake while in the building.

Group E—Educational: This occupancy is intended only where there are six or more children that require adult supervision during an emergency. Education facilities through the 12th grade are anticipated to have trained faculty and staff who know through fire drills the proper procedures to ensure life safety for all school occupants. Like assembly occupancies panic devices are required where there are 50 or more occupants.

Group F—Factory: Factories with moderate and low hazard materials and standard industrial use share many of the requirements for Group B—Business occupancy.

Group H—Hazardous: Facilities that involve the manufacturing, processing, generation, or storage of materials that constitute a physical or health hazard such as explosive, combustible, corrosive, or toxic materials have special limitations and requirements.

Group I—Institutional: These are facilities that house people day and night who are supervised 24 hours a day, seven days a week, either due to medical problems or due to incarceration for the public welfare such as hospitals, nursing homes, nurseries, sanitariums, jails, prisons, and reform schools. Some day-care facilities are also considered institutional occupancies.

Group M—Mercantile: Facilities for display and sale of merchandise presently share the same requirements as Group B—Business.

Group R—Residential: These are facilities used for sleeping other than institutional facilities including both transient facilities such as hotels (R-1) and primarily permanent facilities such as apartment houses and dormitories (R-2), and also one- and two-family dwelling units (R-3).

Group R occupancies may well be the least safe occupancy as the occupants can be sleeping and there is no 24-hour supervision. A sleeping person could inhale sufficient smoke to suffer "smoke inhalation syndrome," which relates to a loss of oxygen supply to the brain that causes sufficient confusion to prevent the person from being able to open a door with a key. More people die in home fires every year than in any other occupancy.

Group S—Storage: These are facilities used for storage including garages, aircraft hangers, and helistops. People may work in these facilities.

Group U—Utility: These facilities are not intended for human occupancy, including private garages, carports, sheds, fences, stables, and towers.

TYPE OF CONSTRUCTION

Another key factor in determining the relative safety of a building is the type of construction. Two specific characteristics are considered—first, whether the structural materials are classified as noncombustible or combustible, and second, how long the structure can be anticipated to not collapse after a fire begins based on hours, for example, 4, 3, 2, or 1-hour fire-rated construction.

Noncombustible Construction

Noncombustible construction requires structural components to be made of steel, concrete, masonry, or some other "limited combustible" materials, but not wood. Noncombustible types of construction may or may not have fire ratings for structural components. Both the IBC and NFPA 5000 consider Type I and Type II construction noncombustible.

Combustible Construction

Combustible construction includes facilities with structural components of wood or other materials determined to be combustible such as some plastics. Combustible types of construction may or may not have fire ratings for structural components. Both model codes consider Type III, Type IV, and Type V construction combustible.

Just as important to the type of construction as whether or not the primary structural materials are combustible is the amount of time the structural components can be expected to continue to support their loads without collapse after a fire starts.

Fire-Rated Construction

The International Building Code requires their highest rated structure, Type IA, to be able to support their design loads for not less than three hours after a fire begins based on either experience or testing. NFPA 5000 requires a four-hour rating for structural components in their highest-rated structure Classified as Type I (442). NFPA 5000's Type I (332) construction would be closer in type of construction to IBC Type IA. These noncombustible buildings are often allowed unlimited height and area because the fire personnel can fight the fire from inside the building without the danger of collapse.

Without becoming too detailed with code nomenclature, both the model codes have classifications for two-hour fire-rated construction, one-hour fire-rated construction, and non-fire rated construction. A non-rated building could collapse in less than one hour after a fire began, which would limit the height and area of the building available to allow fire personnel to fight the fire from the exterior only. Ideally one-hour fire-rated construction is also limited in height and area to allow fighting the fire from outside.

Even the safest occupancy, Group B—Business, is limited to no more than five stories for one-hour fire-rated and non-fire-rated construction, and six stories where an approved automatic sprinkler system is provided.

BUILDING VOLUME—AREA AND HEIGHT

Building area and height is determined by the type of construction along with the occupancy based on Table 503 in the International Building Code and Table 7.4.1 in the NFPA 5000. Unlimited heights and areas are typical for noncombustible construction fire-rated at three hours or higher, with the primary exceptions relating to Hazardous Occupancies. Two-hour fire-rated noncombustible construction may include high-rise buildings, 75 feet or more to the highest occupied floor, with some exceptions.

Height restrictions are given in the tables based on both the number of stories allowed and on the maximum height of the building in feet. Both height limitations differ in the two codes.

The allowable areas in the two model codes appear to be the same to the extent the two are comparable. Moreover, the areas listed in the model codes are just the beginning as area increases are allowed for various reasons.

First, both codes allow area increases based on the percentage of accessible building frontage, because the greater the frontage, the greater the ability for fire personnel to fight the fire from more than one direction. Second, they allow area increases based on having an automatic sprinkler system, and unlike earlier codes, the increase is permitted even if used for height or story increases. Third, they allow area increases based on having more than one story.

FIRE

Collapse

One primary life safety concern during a fire is that the building or a part of the building may collapse during a fire. If a fire starts on the first floor of a high rise building, and it can be estimated to take several hours to evacuate the building and extinguish the fire, it is important that the structural components be capable of continuing to support the building during the evacuation of the building and while firefighters are extinguishing the fire.

Structural components are rated based on the length of time in hours they will continue to be structurally sound during a fire. Tests and experience establish typical fire ratings of one hour, two hours, three hours, and four hours. Heavy timber construction (Type IV) is an example of experience establishing a fire rating. Heavy timber is automatically considered one-hour fire rated. Heavy timber is defined as wood structural members with their smallest dimension nominal six inches or greater.

Containment and Compartmentation

The spread of fire in a structure also provides a means for assuring the safety of the occupants and provides fire personnel time to extinguish a fire. By providing a fire separation proven by test or experience to be able to slow the spread of fire, occupants can be assured safe egress, even if their path goes through a floor with an active fire. Firefighters can also establish staging areas from safe locations knowing they will be protected for at least a minimal amount of time.

Walls, partitions, shafts, floor-ceilings, and roof-ceilings are rated based on the same time ratings as structural components, one hour, two hours, three hours, and four hours. The key difference is that structural components are tested on to ensure the component maintains structural integrity where containment and compartmentation requires openings in the fire-rated construction such as doors, windows, and penetrations, to be fire rated.

In some cases, building components can be fire rated for both structural and containment reasons. The fire rating of a floor assembly in a high rise building would have to meet both requirements. An exterior wall located on or near the property line might also require both structural fire ratings to prevent collapse and containment fire ratings for protection from a fire in an adjacent building. An exterior wall facing a public way, however, would only require a structural fire rating, allowing for limitless unprotected glazed openings facing the public way.

Flame Spread

Wainscoting, paneling, or other finish applied structurally or for decoration, acoustical correction, surface insulation, or similar purposes and used for wall or ceiling finishes are required to be tested for flame spread and for smoke developed.

Flame Spread ratings are measured according ASTM E 84, also referred to as UL 992 and NFPA 255. But whatever is used to describe the test method, it is the "Steiner Tunnel Test."

Classifications:

- Class A—Flame Spread 0 to 25. (Former building codes referred to this as Class I.)
- Class B—Flame Spread 26 to 75. (Former building codes referred to this as Class II.)

- Class C—Flame Spread 76 to 200. (Former building codes referred to this as Class III.)

The International Building Code describes the flame spread requirements in Table 803.5. NFPA 5000 lists flame spread requirements as part of the requirements for each occupancy. The following classifications are general and can vary based on the Occupant Group and on whether or not the facility is sprinklered.

- Enclosed Vertical Exitways (Stairs): Class A.
- Other Exitways (such as Corridors): Class B.
- Rooms in Assembly and Institutional Occupancies: Class B.
- Rooms in other Occupancies: Class C.
- Dwelling Units (R-3): Class C.

Smoke Developed/Smoke Development

ASTM E 84, et al. also tests for the amount of smoke developed during the flame spread test. Smoke developed shall be maximum 450 in all cases. Some of the older model building codes referred to this as "smoke density."

Floor Finishes

It's important to recognize that flame spread and smoke developed classifications are for wall and ceiling finishes only, not for floor finishes. Floor finishes are tested using different testing criteria, ASTM D 2859 or NFPA 253, the flooring radiant panel test. Both codes list Class I and Class II ratings with Class I being more resistant to spreading fire.

Typically Class I floor finish ratings are required in Institutional Occupancy vertical exits, exit passageways, and exit access corridors. Class II floor finish ratings are required in the same locations in other occupancies.

EGRESS—EXIT ACCESS, EXIT, EXIT DISCHARGE

"Means of Egress" is defined as a continuous and unobstructed path of travel from any occupied portion of a facility to a public way. Egress consists of three specifically defined parts: exit access, exit, and exit discharge.

Some egress components can be found in all three parts. A door within an office suite would be part of the exit access. A door into a fire stair would be part of the exit. And a door leading from an egress court to a street would be part of the exit discharge. Stairs, railings, and ramps could also be found in any part of the means of egress.

Comparisons between the International Building Code (IBC) and NFPA 5000 become more difficult related to egress as the earlier model codes used to create the IBC had basic requirements with exceptions and special requirements for various occupancies in Chapter 10. NFPA 5000 is based largely on the NFPA 101 Life Safety Code, where there are some general requirements, but specific requirements are located as part of each occupancy in various chapters.

Occupancy Loads

The first step in designing egress for a facility is to determine the occupant load. Occupant loads are determined based on tables in the model codes, Table 1004.1.2 in the IBC and Table 11.3.1.2 in NFPA 5000. The tables list various occupancies or uses along with a numeric floor area to be used to determine the occupant load.

Both model codes indicate "Business" as allowing for no more than 100 square feet per occupant. Thus the occupant load factor for an architect's office with 3,000 square feet would be 30 occupants, 3,000 divided by 100. The square footage for business occupancies is required to be calculated based on gross square footage. Occupancies such as assemblies and educational classrooms allow calculations to be based on net floor area.

In order to determine egress occupancy loads, it is important to calculate the loads for each defined area on a floor and then to determine the egress occupancy loads for the entire floor.

This can prove difficult in buildings such as speculative office buildings. Calculations would have to be based on anticipated occupancies, the gross area of the floor divided by 100. When the tenant floor plans are established, the resultant calculations will undoubtedly differ because every office area not designed on a 100-square foot division would result in additional occupants.

Thus, a 3,000-square foot floor would be designed with 30 occupants anticipated. But, if the floor were designed with 20 offices with a gross area of 110 square feet each, the total occupants from those offices alone would be 40 occupants. The remaining 800 gross square feet on the floor (3,000 minus 2,200) would result in an additional 8 occupants or more, depending on the configuration, and a total occupant load factor of 48 or more, more than a 60 percent increase.

Changing one or more of the offices (Business) to conference rooms (Assembly) would have an additional increase in occupant load. Changing the whole floor to a restaurant (Assembly) would result in an occupant load factor of 150 or more depending upon the net square feet used without fixed seats, bench, or booth seating, and how much of the floor would be the kitchen area.

Egress Width

Once the occupant load has been determined, it is necessary to compute the egress width required. The total exit width is calculated by multiplying the occupant load factor by a decimal number based on the occupancy type with the result being divided by the required number of exits.

Thus, in the IBC, a stairway serving 1,000 persons in a Business Occupancy in a building without sprinklers would use the factor of 0.3 from Table 1005.1 resulting in $1,000 \times 0.3 = 300$ inches. As the model codes require not less than three separate exits located remotely from each other for occupant loads of 501 to 1,000, the 300 inches would be divided by three, resulting in a requirement that each of the three exit stairs be 100 inches wide.

The same calculation for doors would be based on a factor of 0.2 from IBC Table 1005.1, thus a total of 200 inches for the three exit doors, requiring each door to be not less than 66.67 inches wide. As exit doors are required to have not less than 32″ clear width and not more than 48″ nominal width, the design for the three doors would most likely result in three pairs of 36″ wide doors.

It is important to note that there are major differences in the width factors in the IBC and NFPA 5000.

Exit Access

Exit access starts where an occupant is within a facility and continues until the occupant reaches the "Exit." In a restaurant, it begins at any given seat. In an office, it begins in any office or other occupied part of the office. Exit access requires consideration of the number of exits required for each space, the required separation of exits when more than one exit is required, travel distance to the closest exit, egress through intervening spaces, and corridors.

Number of Exits

When discussing the number of exits from a building, it is typically easiest to assume that each building and each floor in each building requires not less than two exits; each building and each floor in each building with an occupant load of 501 to 1,000 requires three exits; and each building and each floor in each building with an occupant load of more than 1,001 requires at least four exits. Additional exits may be required based on travel distance on each floor to an exit.

Frequently a space within a floor of building might also require two or more exits. The IBC requires any space in occupancies other than H and S to have a second means of egress when the occupant load exceeds 50. Thus, assembly occupancies, such as conference rooms with tables and chairs, in a space of 750 square feet or more would require two means of egress (50 times 15). Similarly office occupancies with 5,000 square feet or more would require two means of egress (50 times 1,000). Although the concept is the same as in earlier model codes, the numbers have changed over the years.

Separation of Exits

Another requirement typical in the model codes is that when more than one exit is required, the exits are required to be separated by a specific distance. Where two exits are required, the separation is required to be not less than one half the maximum diagonal of the space or floor requiring two exits. Where more than two exits are required, the additional exits are required to be located a reasonable distance apart so the others will be available if one becomes blocked.

Travel Distance

Measuring from the most remote point to the closest exit, the typical maximum travel distance is 200 feet in buildings without a sprinkler system and 250 feet in buildings with a sprinkler system. Business occupancies allow up to 300 feet in buildings with sprinklers. Hazardous occupancies and institutional occupancies typically allow less travel distance; and some Factory, Storage, and Utility occupancies allow 300 feet typically and 400 feet in buildings with sprinklers.

Egress through Intervening Spaces

Travel to an exit often requires leaving one room and passing through intervening spaces before reaching an exit. The model codes limit the types of spaces allowed. Typically the IBC allows exiting through adjoining rooms or areas accessory to the area served as long as they provide a discernible path to an exit.

Types of spaces not allowed to intervene with an exit path include kitchens, storage rooms, closets, toilet rooms, and bathrooms, although kitchens in dwelling units are not prohibited. Separate dwelling units and separate tenant spaces may not be part of the required path of travel to an exit.

Corridors

The final segment of an exit access is often a corridor leading to the exit. The standard corridor width is 44 inches, although specific areas have different requirements; for example, Educational occupancies serving 100 or more require 72-inch wide corridors.

A "dead end" is where a corridor continues beyond an exit but does not continue to another exit. The maximum dead end is 20 feet. This means no person would have to travel more than 20 feet in smoke before reaching an end

and then turning around and having to retrace their steps to the corridor. There were no exceptions in the older model building codes, but both the IBC and NFPA 5000 have a number of exceptions, including allowing Business occupancies with sprinkler systems a dead end as long as 50 feet.

Fire-rated corridors are required in most occupancies without sprinkler systems, as well as in some Institutional and in all Hazardous occupancies even with sprinklers. Fire-rated corridors are typically required in unsprinklered buildings where the occupant load exceeds 30 and in unsprinklered Residential occupancies where the occupant load exceeds 10. Even with a sprinkler system, Residential occupancies with an occupant load greater than 10 require a 0.5 hour fire-resistive rating.

Finally, fire-resistant-rated corridors are required to be continuous from the point of entry to an exit with no intervening rooms except foyers, lobbies, and reception rooms constructed with the same fire-resistant rating as the corridor.

Exits

The exit is that part of egress that either takes the occupant from the building or floor to a vertical exit enclosure (stair), an exit passageway, a horizontal exit, or an exterior exit ramp or stairway.

Vertical exit enclosures are interior exit stairways and interior exit ramps. Typically vertical exit enclosures are required to be 2-hour fire rated when serving four or more stories including basements (but not mezzanines). Where serving fewer than four stories they are required to be 1-hour fire rated. In rare instances, vertical exit enclosures can be non-fire rated such as in a single residential dwelling unit, but care should be taken to verify the applicability of the exception.

Exit passageways serve to connect multiple vertical exit enclosures where the stairs cannot be continuous. An example would be the passageway connecting an enclosed stairway in a tower to another stairway at the perimeter of the building pedestal. An exit passageway could also serve to connect an enclosed stairway to an exterior wall where the stair walls are not at the building perimeter. The important aspect of the exit passageway is to realize it must comply with the same basic requirements as the vertical exit enclosure, meaning it may not have toilet rooms, mechanical closets, or storage closets opening directly into the exit passageway. It is easy to confuse an exit passageway with a corridor that does not have the same restrictions.

Horizontal exits were originally used in Institutional occupancies to allow the moving of persons who were nonambulatory (patients) or persons who were being restrained (prisoners) without having to evacuate them from the building. The process gained acceptance and now typically one exit from a space can be a horizontal exit as long as the other exit is either an enclosed vertical exit or an exit directly to the outside. The horizontal exit divides a building or floor into separate compartments with a 2-hour fire-rated partition separating the floor from exterior wall to exterior wall.

Exterior exit ramps and stairways provide the fourth type of exit as long as the building is not more than six stories or 75 feet in height. The exterior exit ramps and stairways are required to be open on at least one side and are required to be separated from the building by fire rated construction matching that required for vertical exit enclosures.

Exit Discharge

The final segment of the means of egress is the exit discharge. The exit discharge can be as simple as the door leading from an vertical exit

enclosure to the public street and as complex as the path of travel from a theater building at an amusement park to the public way including the path through another building used for selling and collecting entry tickets. Open areas connecting the exit door to the public way may require conformance with requirements for egress courts.

ACCESS FOR PERSONS WITH DISABILITIES—CODE VS. ACT OF DISCRIMINATION

Present model codes include specific requirements for ensuring access for persons with disabilities based on scoping requirements in the model code along with specific accessibility requirements in the International Code Council (ICC) and American National Standards Institute (ANSI) A117.1 and the American National Standard for *Accessible and Usable Buildings and Facilities*.

The scoping requirements in the model code provide requirements for accessible routes and entrances; parking and passenger loading facilities; dwelling and sleeping units; special occupancies such as accessible assembly area seating; toilets and bathing facilities; kitchens; drinking fountains; elevators; lifts; storage; service facilities such as check-out aisles; controls; operating mechanisms; and hardware; recreational facilities; and signage.

These scoping requirements typically tell when these design elements are required to meet the specific accessibility requirements in ICC/ANSI A117.1.

It is important to recognize that the requirements in the model building codes are the same as the requirements for height and area limitations, type of construction, fire ratings, egress, and general construction. Where these chapters are adopted by state or local authorities, the building officials have both the right and the responsibility to enforce compliance. Citizens might have rights to disagree on the building official's interpretation, based on the actual wording of the state or local law, but they cannot bring a lawsuit against the property owner or design team based on the failure to comply with the code unless there is a specific law permitting such a complaint (as there is in California). The Fair Housing Act and the Americans with Disabilities Act are not building codes and are subject to totally different enforcement processes.

FEDERAL STANDARDS AND REQUIREMENTS: OSHA, FHAG, ADAAG, UFAS

The U.S. government has developed several standards and requirements that impact building design and construction. Best known perhaps is the Occupational Safety and Health Act or OSHA. In addition, there are the Safety Glass standard 16 CPR 1201, the Fair Housing Act, and the Americans with Disabilities Act; federal agencies often adopt specific standards and requirements for construction of buildings and spaces for their own use or under their jurisdiction.

OSHA and the related CalOSHA in California provide specific requirements for the workplace that include stairs, railings, ladders, and protection of openings.

The safety glass standard still exists but the basic requirements have long ago been adopted into the model building codes, which limit those interested in the federal standards to lawyers and those trying to prove a violation-caused injury.

The Fair Housing Act and Americans with Disabilities Act are new types of federal regulation. They are similar to the building code requirements for access for persons with disabilities but they are part of the nondiscrimination laws rather than a federal building code. This means that enforcement is done through the federal courts system based on complaints by persons who believe they have been discriminated against based on their disability.

The results are three complete sets of federal requirements for accessibility—FHAG, ADAAG, and UFAS. The Fair Housing Accessibility Guidelines (FHAG) provide requirements to ensure nondiscrimination in privately funded housing, The Americans with Disabilities Act Accessibility Guidelines (ADAAG) provide requirements to ensure nondiscrimination in facilities other than privately funded housing, including publicly funded housing. And, the Uniform Federal Accessibility Standards (UFAS) provides requirements for construction under four specific federal government agencies including the United States Postal Service (USPS), the Department of Defense (DOD), and the General Services Administration (GSA).

For those slightly confused about the relationship of UFAS and the Americans with Disabilities Act, UFAS is solely for federally funded projects, but is an option to the ADA Access Guidelines for state and local government projects. Only the ADA Accessibility Guidelines (ADAAG) defines access requirements to privately owned facilities.

STATE BUILDING CODES

In an attempt to provide uniformity throughout their communities, many states have adopted one of the model codes and mandated its use throughout the state. Adoption of a particular model code by a state does provide some uniformity; however, it is possible for both the state and a community within the state to adopt revisions and additions to the adopted model code. Although these additions and revisions should not conflict (the operative words are *should not*), they can still cause a major amount of confusion for architects.

Most states adopt one of the model code families with little revision. States such as California and Florida have extensive differences, causing their state codes to vary extensively from the model codes they were originally based upon.

CITY BUILDING CODES

Although states may mandate adoption of a specific model building code, the enforcement of the code is typically left to building officials in each community. Sometimes local conditions mandate the need for local modifications.

Even when there is no state mandated model building code, most communities adopt one of the model code families with little revision except to the administrative chapters that describe the operations of the local building departments.

Some larger cities are in the habit of adopting revisions to the model codes, and in some cases, they can totally rewrite the model code to their own requirements. Some cities have such extensive differences that their codes are sometimes not even similar to the model codes they were originally based upon.

PLANNING AND ZONING CODES

Planning and zoning codes are usually defined by specific local interests and conditions. The City and County of Saint Louis, Missouri, limits buildings in the downtown area to heights lower than their famous Gateway Arch. Rural communities with no sewer systems limit the number of buildings based on the amount of area needed for the septic system either as a function of how many acres are required per dwelling unit, or in some cases how many acres are required per bathroom. Because planning and zoning codes are not nationalized, architects must obtain them from the local officials and ascertain their impact to the project as early as possible.

Planning and zoning codes can restrict specific areas in the community to certain types of building occupancies (commercial, residential, light industrial, etc.). They can also restrict building types in specific areas (construction materials such as concrete, steel, masonry, or wood frame construction). They can limit the heights and allowable square footage of buildings within a particular area of the community, sometimes based on the type of fire fighting equipment available. And, they can require open space, limit shadowing, and even place design requirements such as building (or glass) colors.

The original purpose of the planning and zoning codes was to limit construction of highly combustible wood frame construction in cities that had gone through major fires around the turn of the century, such as Chicago and St. Louis, and the site of the great fire that followed the San Francisco earthquake of 1906. Later, these planning and zoning codes were designed to prevent factories causing pollution from being built in the residential and commercial areas of a community. They also were intended to prevent construction on flood plains, seismic faults, and hillsides subject to landslides. At present it can sometimes be difficult to recognize the original purposes.

FIRE CODES AND FIRE ZONES

Fire code requirements are intended to provide for public health and safety related to management and use of existing structures in case of an emergency. Model fire codes are generally written with requirements for management rather than construction. But fire codes can provide design information regarding the locations, types, and sizes of fire extinguishers, sprinklers, and wet and dry standpipe systems.

Standpipe systems direct water to upper floors of a building to provide water to firefighters in a particular area without having to bring external hoses up the stairs or into interior building spaces. A dry standpipe has no water in it until the connection is made at an outside hose connection. A wet standpipe has water in it at all times. Sprinklers also have water or some fire suppressing chemical in the pipes at all times.

Fire zones are often established to recognize the type of firefighting equipment and trained personnel available in a particular area. Where a particular fire station is not designed for a hook-and-ladder truck, the height of buildings in that station's area may be limited to two or three stories so the fires could be fought from the ground and standard ladders. Fire zones are sometimes incorporated into the building code.

ICC/ANSI A117.1 ACCESSIBLE AND USABLE BUILDINGS AND FACILITIES

The ICC/ANSI A117.1 standards are not in and of themselves building regulations. However, when state and local authorities adopt the model building code including the chapters on accessibility they adopt the ICC/ANSI A117.1 standards as part of their regulatory authority.

Even when the model code chapters on accessibility are not adopted familiarity with the standards published in ICC/ANSI A117.1 can help architects make their buildings and facilities more accessible.

Although the various regulations and standards relating to accessibility may never be identical they do have a common purpose. Whether the enforcement is by a local authority or by a complaint by an aggrieved person, architects need to develop an understanding of more than the words and measurements involved in accessibility.

CONCLUSION

Public health, safety, and welfare are the responsibility of every architect. Public agencies and authorities should be seen as members of the design team interested in helping architects achieve the degree of safety necessary for that responsibility. They should not be viewed as outsiders nor as people who can take on architectural design responsibility.

Architects need to understand the concepts of the building codes, incorporate the codes into their designs, and be diligent in making sure the finished project is in compliance.

When starting a project, always look for compliance with the following:

- Occupancy/Use
  (A, B, E, F, H, I, M, R, S, U)
- Construction Type (I, II, III, IV, V)
- Allowable Area
- Allowable Height
- Separations and Shafts
- Fire Resistance
- Fire-Extinguishing Systems
- Means of Egress
- Access for Persons with Disabilities
- Engineering and Material Requirements

LESSON 8 QUIZ

1. Building codes are intended to protect which of the following?

    A. Public health

    B. Public welfare

    C. Public safety

    D. All of the above

2. What model building code is presently published in the United States?

    A. International Building Code

    B. Uniform Building Code

    C. National Building Code

    D. Standard Building Code

3. Which can be considered as the safety occupancy in the model building codes?

    A. Assembly

    B. Business

    C. Commercial

    D. Residential

4. Combustible types of construction include all but which of the following?

    A. Type II N

    B. Type III N

    C. Type V N

    D. Type V one hour

5. Which of the following factors are to be considered in determining the allowable area of a building?

    I.   Occupancy

    II. Type of Construction

    III. Percentage of building frontage accessible to a public way or open space

    IV. Automatic sprinkler system

    V.   Number of stories of building

    A. I, II, and IV

    B. I, II, III, and IV

    C. I, II, IV, and V

    D. I, II, III, IV, and V

6. Fire ratings including one hour, two hours, three hours, and four hours could be related to which of the following?

    A. The ability of a wide flange structural steel member with fireproofing to withstand fire without collapse for the designated time

    B. The ability of an exterior curtain wall to withstand fire for up to the designated time without collapse

    C. The ability of an enclosed vertical exit stair to maintain a safe passage for persons exiting the building without being exposed to fire for the designated time

    D. The ability of a door to prevent the spread of fire through the door opening for the designated time

7. What is the purpose of the Steiner Tunnel Test?

   A. To test waterproofing for tunnel construction under bodies of water

   B. To test the flame spread rating of finishes for interior walls and ceilings

   C. To test the fire ratings of interior finishes

   D. To test the wind loads on a curtain wall

8. Which of the following are considered the primary components of the "Means of Egress"?

   I.   Exit access

   II. Exit door

   III. Exit

   IV. Exit corridor

   V.   Exit discharge

   A. I, II, IV, and V

   B. II and IV

   C. I, III, and V

   D. I, II, III, IV, and V

9. The need for three exits can be based on which of the following?

   I.   Occupancy type

   II.   Type of construction

   III. Number of occupants (more than 500 but not more than 1,000)

   IV.   Travel distance to an exit

   V.   Intervening rooms

   A. I and II   C. III and V

   B. III and IV   D. IV and V

10. Which of the following provide information for designing facilities accessible to persons with disabilities?

   I.   ICC/ANSI A117.1

   II.   FHAG

   III. ADAAG

   IV.   UFAS

   V.   IBC

   A. I, II, and III

   B. I, II, III, and IV

   C. II, III, and IV

   D. I, II, III, IV, and V

SUSTAINABLE DESIGN

by Jonathan Boyer, AIA

INTRODUCTION

Architects can no longer assume that buildings function independently of the environment in which they are placed. In the late 1800s, the machine age offered the lure of buildings that were self-sufficient and independent of their natural surroundings—"The Machine for Living," as LeCorbusier proclaimed.

In the middle of the 20th century, the promise of endless and inexpensive nuclear energy lured architects into temporarily ignoring the reality of the natural elements affecting architectural design. Why worry about natural systems if energy was going to be infinite and inexpensive? Glass houses proliferated.

Energy is not free, the global climate is changing, and the viability of natural ecosystems is diminishing. Architects are designing structures that affect all these natural ecosystems. Much as Marcus Vitruvius wrote thousands of years ago that architects must be sensitive to the local environment, architects are returning to study the virtues of tuning to natural systems. Contemporary architects must combine their knowledge of the benefits of natural systems with the understanding of the selective virtues of contemporary innovative technologies.

This lesson focuses on the fundamental principles of environmental design that have evolved over the thousands of years that humans have been creating spatial solutions.

HISTORY OF SUSTAINABLE DESIGN

In early human history, builders of human habitats used materials that occurred naturally in the earth, such as stone, wood, mud, adobe bricks, and grasses. With nomadic tribes and early civilizations, the built environment made little impact on the balance of natural elements. When abandoned, the grass roof, adobe brick, or timber beam would slowly disintegrate and return to the natural ecosystem. Small human populations and the use of natural materials had very little impact on a balanced natural ecosystem.

But as human populations expanded and settlements moved into more demanding climates, natural materials were altered to become more durable and less natural. In fact, archeological finds demonstrate some of the human creations

that are not easily recycled into the earth; fired clay, smelted ore for jewelry, and tools are examples of designs that will not easily reintegrate into the natural ecosystem. These materials may be reprocessed (by grinding, melting, or reworking) into other human creations, but they will never be natural materials again.

As human populations expanded, there is strong evidence that some civilizations outgrew their natural ecosystem. When overused, land became less fertile and less able to support crops, timber, and domesticated animals necessary for human life. The ancient solution was to move to a more desirable location and use new natural resources in the new location, abandoning the ecologically ruined home site.

The realization that global natural resources are limited is an age-old concept. The term *conservation,* which came into existence in the late 19th century, referred to the economic management of natural resources such as fish, timber, topsoil, minerals, and game. In the United States, at the beginning of the 20th century, President Theodore Roosevelt and his chief forester, Gifford Pinchot, introduced the concept of conservation as a philosophy of natural resource management. The impetus of this movement created several pieces of natural legislation to promote conservation and increased appreciation of America's natural resources and monuments.

In the middle of the 1960s, Rachel Carson published *Silent Spring,* a literary alarm that revealed the reality of an emerging ecological disaster—the gross misunderstanding of the value and hazards of pesticides. The pesticide DDT and its impact on the entire natural ecosystem was dramatic; clearly, some human inventions were destructive and could spread harm throughout the ecosystem with alarming speed and virulence. Birds in North America died from DDT used to control malaria in Africa. Human creations were influenced by the necessities of the natural cycles of the ecosystem. Human toxic efforts could no longer be absorbed by the cycles of nature. Human activities became so pervasive and potentially intrusive that there needed to be a higher level of worldwide ecological understanding of the risk of disrupting the ecosystem.

Architects, as designers of the built environment, realize the ecological impact of their choices of architectural components, such as site selection, landscaping, infrastructure, building materials, and mechanical systems. The philosophy of sustainable design encourages a new, more environmentally sensitive approach to architectural design and construction.

There are many credos for the approach to a new sustainable design. Some architectural historians maintain that the best architects (Vitruvius, Ruskin, Wright, Alexander) have always discussed design in terms of empathy with nature and the natural systems. Now it is evident that all architects should include the principles of sustainable design as part of their palette of architectural best practices.

PRINCIPLES OF SUSTAINABLE DESIGN

The tenets outlined below indicate why it is necessary to maintain the delicate balance of natural ecosystems.

1. In the earth's ecosystem (the area of the earth's crust and atmosphere approximately five miles high and five miles deep), there is a finite amount of natural resources. People have become dependent on elements such as fresh water, timber, plants, soil, and ore, which are processed into necessary pieces of the human environment.

2. Given the laws of thermodynamics, energy cannot be created or destroyed. The resources that have been allotted to manage existence are contained in the ecosystem.

3. All forms of energy tend to seek equilibrium and therefore disperse. For example, water falls from the sky, settles on plants, and then percolates into the soil to reach the subterranean aquifer. Toxic liquids, released by humans and exposed to the soil, will equally disperse and eventually reach the same underground reservoir. The fresh water aquifer, now contaminated, is no longer a useful natural resource.

There is a need to focus on the preservation of beneficial natural elements and diminish or extinguish natural resources contaminated with toxins and destructive human practices.

There are many credos for environmental responsibility. One, *The Natural Step,* was organized by scientists, designers, and environmentalists in 1996.

They were concerned with the preservation of the thin layer that supports human life in a small zone on the earth's surface: the ecosphere (five miles of the earth's crust) and the biosphere (five miles into the troposphere of the atmosphere).

Their principles are as summarized as follows:

■ Substance from the earth's crust must not systemically increase in the ecosphere.

Elements from the earth such as fossil fuel, ores, timber, etc., must not be extracted from the earth at a greater rate than they can be replenished.

■ Substances that are manufactured must not systemically increase in the ecosphere.

Manufactured materials cannot be produced at a faster rate than they can be integrated back into nature.

■ The productivity and diversity of nature must not be systemically diminished.

This means that people must protect and preserve the variety of living organisms that now exist.

■ In recognition of the first three conditions, there must be a fair and efficient use of resources to meet human needs.

This means that human needs must be met in the most environmentally sensitive way possible.

■ Buildings consume at least 40 percent of the world's energy. Thus, they account for about a third of the world's emissions of heat-trapping carbon dioxide from fossil fuel burning, and two-fifths of acid rain-causing carbon dioxide and nitrogen oxides.*

The built environment has a monumental impact on the use of materials and fuels to create shelter for human beings. The decisions about the amount and type of materials and systems that are employed in the building process have an enormous impact on the future use of natural resources. Architects can affect and guide those decisions of design to influence the needs of sustainability and environmental sensitivity.

---

* *Sources: David Malin Roodman and Nicholas Lessen, "Building Revolution: How Ecology and Health Concerns are Transforming Construction." Worldwatch Paper 124 (Washington DC, Worldwatch Institute, 1995); Sandra Mendler & William Odell, The HOK Guidebook to Sustainable Design, (New York: John Wiley & Sons, Inc., 2000).*

SUSTAINABLE SITE PLANNING AND DESIGN

Most architectural projects involve the understanding of the design within the context of the larger scale neighborhood, community, or urban area in which the project is placed.

If the building will be influenced by sustainable design principles, its context and site should be equally sensitive to environmental planning principles.

Sustainable design encourages a re-examination of the principles of planning to include a more environmentally sensitive approach. Whether it is called Smart Grow, sustainable design, or environmentally sensitive development practice, these planning approaches have several principles in common.

Site Selection

The selection of a site is influenced by many factors including cost, adjacency to utilities, transportation, building type, zoning, and neighborhood compatibility. In addition to these factors, there are sustainable design standards that should be added to the matrix of site selection decisions:

■ **Adjacency to public transportation**
If possible, projects that allow residents or employees access to public transportation are preferred. Allowing the building occupants the option of traveling by public transit may decrease the parking requirements, increase the pool of potential employees, and remove the stress and expense of commuting by car.

■ **Flood plains**
In general, local and national governments hope to remove buildings from the level of the 100-year floodplain. This can be accomplished by either raising the building

at least one foot above the 100-year elevation or locating the project entirely out of the 100-year floodplain.

This approach reduces the possibility of damage from flood waters, and possible damage to downstream structures hit by the overfilled capacity of the floodplain.

- **Erosion, fire, and landslides**
  Some ecosystems are naturally prone to fire and erosion cycles. Areas such as high slope, chaparral ecologies are prone to fires and mud slides. Building in such zones is hazardous and damaging to the ecosystem and should be avoided.

- **Sites with high slope or agricultural use**
  Sites with high slopes are difficult building sites and may disturb ecosystems, which may lead to erosion and topsoil loss. Similarly, sites with fertile topsoil conditions—prime agricultural sites—should be preserved for crops, wildlife, and plant material, not building development.

- **Solar orientation, wind patterns**
  Orienting the building with the long axis generally east west and fenestration primarily facing south may have a strong impact on solar harvesting potential. In addition, protecting the building with earth forms and tree lines may reduce the heat loss in the winter and diminish summer heat gain.

- **Landscape site conditions**
  The location of dense, coniferous trees on the elevation against the prevailing wind (usually west or northwest) may decrease heat loss due to infiltration and wind chill factor. Sites with deciduous shade trees can reduce summer solar gain if positioned properly on the south and west elevations of the buildings.

Alternative Transportation

Sites that are near facilities that allow several transportation options should be encouraged. Alternate transportation includes public transportation (trains, buses, and vans); bicycling amenities (bike paths, shelters, ramps, and overpasses); carpool opportunities that may also connect with mass transit; and provisions for alternate, more environmentally sensitive fuel options such as electricity or hydrogen.

Reduction of Site Disturbance

Site selection should conserve natural areas, and restore wildlife habitat and ecologically damaged areas. In some areas of the United States, less than 2 percent of the original vegetation remains. Natural areas provide a visual and physical barrier between high activity zones. Additionally, these natural areas are aesthetic and psychological refuges for humans and wildlife.

Storm Water Management

There are several ways by which reduced disruption of natural water courses (rivers, streams and natural drainage swales) may be achieved:

- Provide on-site infiltration of contaminants (especially petrochemicals) from entering the main waterways. Drainage designs that use swales filled with wetland vegetation is a natural filtration technique especially useful in parking and large grass areas.

- Reduce impermeable surface and allow local aquifer recharge instead of runoff to waterways.

- Encourage groundwater recharge.

Ecologically Sensitive Landscaping

The selection of indigenous plant material, contouring the land, and proper positioning of shade trees may have a positive effect on the landscape appearance, maintenance cost, and ecological balance. The following are some basic sustainable landscape techniques:

■ Install indigenous plant material, which is usually less expensive, to ensure durability (being originally intended for that climate) and lower maintenance (usually less watering and fertilizer).

■ Locate shade trees and plants over dark surfaces to reduce the "heat island effect" of surfaces (such as parking lots, cars, walkways) that will otherwise absorb direct solar radiation and retransmit it to the atmosphere.

■ Replace lawns with natural grasses. Lawns require heavy maintenance including watering, fertilizer, and mowing. Sustainable design encourages indigenous plant material that is aesthetically compelling but far less ecologically disruptive.

■ In dry climates, encourage xeriscaping (plant materials adapted to dry and desert climates); encourage higher efficiency irrigation technologies including drip irrigation, rainwater recapture, and gray water reuse. High efficiency irrigation uses less water because it supplies water directly to the plant's root areas.

Reduction of Light Pollution

Lighting of site conditions, either the buildings or landscaping, should not transgress the property and not shine into the atmosphere. Such practice is wasteful and irritating to the inhabitants of surrounding properties. All site lighting should be directed downward to avoid "light pollution."

Open Space Preservation

The quality of residential and commercial life benefits from opportunities to recreate and experience open-space areas. These parks, wildlife refuges, easements, bike paths, wetlands, or play lots are amenities that are necessary for any development.

In addition to the aforementioned water management principles, the following are principles of design and planning that will help increase open-space preservation:

■ **Promote in-fill development** that is compact and contiguous to existing infrastructure and public transportation opportunities.

In-fill development may take advantage of already disturbed land without impinging on existing natural and agricultural land.

In certain cases, in-fill or redevelopment may take advantage of existing rather than new infrastructure.

■ **Promote development that protects natural resources** and provides buffers between natural and intensive use areas.

First, the natural areas (wetlands, wildlife habitats, water bodies, or flood plains) in the community in which the design is planned should be identified.

Second, the architect and planners should provide a design that protects and enhances the natural areas. The areas may be used partly for recreation, parks, natural habitats, and environmental education.

Third, the design should provide natural buffers (such as woodlands and grasslands) between sensitive natural areas and areas of intense use (factories, commercial districts, housing). These buffers may offer visual, olfactory, and auditory protection between areas of differing intensity.

Fourth, linkages should be provided between natural areas. Isolated islands of natural open space violate habitat boundaries and make the natural zones seem like captive preserves rather than a restoration or preservation of natural conditions.

Fifth, the links between natural areas may be used for walking, hiking, or biking, but should be constructed of permeable and biodegradable material. In addition, the links may augment natural systems such as water flow and drainage, habitat migration patterns, or flood plain conditions.

■ **Establish procedures that ensure the ongoing management of the natural areas** as part of a strategy of sustainable development.

Without human intervention, natural lands are completely sustainable. Cycles of growth and change including destruction by fire, wind, or flood have been occurring for millions of years. The plants and wildlife have adapted to these cycles to create a balanced ecosystem.

Human intervention has changed the balance of the ecosystem. With the relatively recent introduction of nearby human activities, the natural cycle of an ecosystem's growth, destruction, and rebirth is not possible.

Human settlement will not tolerate a fire that destroys thousands of acres only to liberate plant material that reblooms into another natural cycle.

The coexistence of human and natural ecosystems demands a different approach to design. This is the essence of sustainable design practices, a new approach that understands and reflects the needs of both natural and human communities.

AHWAHNEE PRINCIPLES

In 1991, in the Ahwahnee Hotel in Yosemite National Park, a group of architects, planners, and community leaders met to present com-munity principles that express new, sustainable planning ideas. These principles are summarized below.

Preamble

Existing patterns of urban and suburban development seriously impair our quality of life. The symptoms are: more congestion and air pollution resulting from our increased dependence on automobiles, the loss of precious open space, the need for costly improvements to roads and public services, the inequitable distribution of economic resources, and the loss of a sense of community. By drawing upon the best from the past and the present, we can plan communities that will more successfully serve the needs of those who live and work within them. Such planning should adhere to certain fundamental principles.

Community Principles

■ All planning should be in the form of complete and integrated communities containing housing, shops, workplaces, schools, parks, and civic facilities, essential to the daily life of the residents.

■ Community size should be designed so that housing, jobs, daily needs, and other activities are within easy walking distance of each other.

■ As many activities as possible should be located within easy walking distance of transit stops.

■ A community should contain a diversity of housing types to enable citizens from a wide range of economic levels and age groups to live within its boundaries.

■ Businesses within the community should provide a range of job types for the community's residents.

■ The location and character of the community should be consistent with a larger transit network.

- The community should have a central focus that combines commercial, civic, cultural, and recreational uses.

- The community should contain an ample supply of specialized open space in the form of squares, greens, and parks, whose frequent use is encouraged through placement and design.

- Public spaces should be designed to encourage the attention and presence of people at all hours of the day and night.

- Each community or cluster of communities should have a well-defined edge, such as agricultural greenbelts or wildlife corridors, permanently protected from development.

- Streets, pedestrian paths, and bike paths should contribute to a system of fully connected and interesting routes to all destinations. Their design should encourage pedestrian and bicycle use by being small and spatially defined by buildings, trees, and lighting, and by discouraging high speed traffic.

- Wherever possible, the natural terrain, drainage, and vegetation of the community should be preserved with superior examples contained within parks or greenbelts.

- The community design should help conserve resources and minimize waste.

- Communities should provide for the efficient use of water through the use of natural drainage, drought tolerant landscaping, and recycling.

- The street orientation, the placement of buildings, and the use of shading should contribute to the energy efficiency of the community.

Regional Principles

- The regional land-use planning structure should be integrated within a larger transportation network built around transit rather than freeways.

- Regions should be bounded by and provide a continuous system of greenbelt/wildlife corridors to be determined by natural conditions.

- Regional institutions and services (government, stadiums, museums, and so forth) should be located in the urban core.

- Materials and methods of construction should be specific to the region, exhibiting a continuity of history and culture and compatibility with the climate to encourage the development of local character and community identity.

Implementation Principles

- The general plan should be updated to incorporate the above principles.

- Rather than allowing developer-initiated, piecemeal development, local governments should take charge of the planning process. General plans should designate where new growth, in-fill, or redevelopment will be allowed to occur.

- Prior to any development, a specific plan should be prepared based on these planning principles.

- Plans should be developed through an open process and participants in the process should be provided visual models of all planning principles.

Source: Local Government Commission's Center for Livable Communities, http://lgc.org/clc/.

USGBC—U.S. GREEN BUILDING COUNCIL

Incorporated as a nonprofit trade association in 1993, the U.S. Green Building Council (USGBC) was founded with a mission "to promote buildings that are environmentally responsible, profitable and healthy places to live and work." It is formed of leaders from across the building industry who head a national consensus for producing a new generation of buildings that deliver high performance inside and out.

The core of the USGBC's work is the creation of the Leadership in Energy and Environmental Design (LEED) green building rating system. LEED provides a complete framework for assessing building performance and meeting sustainability goals. Based on well-founded scientific standards, LEED emphasizes state of the art strategies for sustainable site development, water savings, energy efficiency, materials selection, and indoor environmental quality. LEED recognizes achievements and promotes expertise in green building through a comprehensive system offering project certification, professional accreditation, training, and practical resources.

USGBC committees are actively collaborating on new and existing LEED standards, including a standard for homes, neighborhood development, and commercial interiors.

Their Web site is: *www.usgbc.org.*

ARCHITECTURAL PROCESS

After the planning process has been concluded, and the site has been selected, the architectural team will begin to focus on the project, including the project's buildings and related infrastructure.

Traditionally, the architect is faced with four components to every design decision: cost, function, aesthetics, and time. The new paradigm adds *sustainability* to this list.

The ingredients of the normal process have been discussed previously, but the new ingredient, sustainability, changes the meaning of all these pieces of the architectural process.

Cost

As architects put together budgets for their clients, they are always concerned with the first costs of the design components—the initial cost to purchase and install the design element.

Sustainable design has made the economic decision process more holistic. The decision to select a design element (such as a window, door, flooring, exterior cladding, or mechanical system) is now concerned with the "life cycle" cost of the design.

Life-Cycle Costing

Life-cycle costing is concerned not only with the first cost, but the operating, maintenance, periodic replacement, and residual value of the design element.

For example, two light fixtures (A and B) might have different first cost: Fixture A has a 10 percent more expensive first cost than B. But when the cost of operation (the lamps use far less energy per lumen output) and the cost of replacement (the bulbs of A last 50 percent longer than the bulbs of Fixture B) is evaluated, Fixture A has a far better life-cycle cost and should be selected.

In this kind of comparison, the life-cycle cost may be persuasive; the extra cost of Fixture A may be recovered in less than two years due to more efficient operation and replacement savings. In this situation, the architect justified Fixture A to the owner, who benefits from a more energy efficient lighting that continues to save the owner operating costs for the life of the building.

Matrix Costing

While designing a typical project, the architect faces numerous alternate decisions, a process that may be both intriguing and complex.

Sustainable design adds an ingredient to the matrix of decisions that may actually help the composition.

For example, decisions that allow the improved efficiency of the building envelope, light fixtures, and equipment may permit the architect to allow the engineer to reduce the size of the HVAC system, resulting in a budgetary trade-off. The extra cost of the improved envelope may be economically balanced by the diminished cost of the mechanical system.

This type of economic analysis, which evaluates cost elements in a broad matrix of interaction, is a very valuable architectural skill. The ability to understand the interaction between different building systems in a creative and organized fashion can differentiate an excellent from a simply adequate architectural design.

Function

Functionality is one of the primary standards of architectural design. If the building does not perform according to the client's needs, then the building design has failed.

Years ago, the design element could perform at the highest level regardless of its impact on the environment or energy use. The fact that many industrial and residential buildings are operating in 2003 much more efficiently than 1960 is evidence that the building design and construction profession is learning how to tune buildings to a higher degree of energy operation. But, with diminishing natural resources and increasing pollution of the environment, even more efficient design is necessary.

Today, architects will include sustainability in the selection of optimal functional design components.

For example, a roof system must be able to withstand a variety of weather conditions, be warranted to be durable a minimum of years, be able to be applied in a range of weather conditions, and have a surface with reflectivity that does not add to the urban heat effect.

Time

The schedule of a project is always a difficult part of the reality of the design process. Time is a constraint that forces a systematic and progressive evaluation of the design components.

The sustainable component of the architectural process may add to the amount of time the architect will spend on the research for the project.

The architect may spend more time on a sustainable design with the result being a more integrated, sustainable project.

Aesthetics

The aesthetic of a project is the combination of the artistry of the architect and the requirements of the project.

Sustainable design has the reputation of emphasizing function and cost over beauty and appeal.

It is the architect's responsibility to keep all the design tools in balance. A project without aesthetic consideration will fail the client, its user, and the potential client who may be deciding between the normal design and one that considers a broader, integrated, sustainable approach.

Sustainability

The fifth point is a new component that leads to a new approach to the design process.

Sustainable designs should have five goals:

1. Use less
2. Recycle components
3. Use easily recycled components
4. Use fully biodegradable components
5. Do not deplete natural resources necessary for the health of future generations

STANDARDS FOR EVALUATION

How can we objectively evaluate the quality of a sustainable project?

The architect is faced with responding to many standards and regulations in the course of assembling a design. Building codes, life safety standards, fire code, zoning regulations, and health and sanitary regulations are some of the many municipal, state, and federal standards that an architect must evaluate in the course of any project.

Sustainability is a new filter for the design process and there are several organizations that have offered checklists for evaluating the inclusion of environmentally sensitive elements into the project.

One of the measures of performance is LEED (Leadership in Energy & Environmental Design), which is sponsored by the USGBC (U.S. Green Building Council). This standard was developed in the 1990s by a consortium of building owners, architects, suppliers, engineers, contractors, and governmental agencies.

The goal of LEED and similar environmental design standards is to introduce new sustainable approaches and technologies to the construction industry. LEED is a voluntary environmental rating system that is organized into six categories:

1. Sustainable sites
2. Water efficiency
3. Energy and atmosphere
4. Materials and resources
5. Indoor air quality
6. Innovation and design practice

LEED covers the range of architectural decisions, including site design, water usage, energy conservation and production, indoor air quality, building materials, natural lighting, views of the outdoors, and innovative design components.

The LEED point award matrix is a mixture of teaching, persuasion, example, and incentive. It is good checklist for the entire project team to evaluate the quality of sustainable design decisions for the complete project—from initial planning through final construction, maintenance, and training procedures.

These categories combine *prerequisites* (basic sustainable practices such as building commissioning, plans for erosion control, or meeting minimum indoor air quality standards) with optional *credits* (water use reduction, heat island reduction, or measures of material recycled content).

Most of the credits are performance based—solutions based on system performances against an established standard such as American Society of Heating, Refrigeration and Air Conditioning Engineers (ASHRAE). ASHRAE has created one of most widely recognized standards of energy design that is used by mechanical engineers and architects.

For example, one credit (under the Energy and Resources category) is "Optimize Energy Performance."

The number of points for this credit depend on how the architectural and engineering team can optimize the design of the building's energy systems against the ASHRAE 90.1 standards.

The possible design solutions include optimizing the heating, cooling, fans, pumps, water, and interior lighting systems.

In the graduated point matrix for a new building, if the team improves the performance (against ASHRAE standards) by 15 percent they receive one point and if they manage to improve by 60 percent they receive ten points.

LEED describes suggested results but allows the architectural team to find a variety of solutions. The LEED certification awards range from Bronze at 40 percent compliance to Platinum at 81 percent compliance. The LEED certification is innovative and rigorous, and currently there are fewer than a half dozen platinum buildings in the United States.

THE SUSTAINABLE DESIGN PROCESS

Is a sustainable design organized and implemented differently from a conventional design?

The Design Team

What kind of design team is necessary for a sustainable project?

The scope of a sustainable design invites an expanded team approach, which may include the following:

- Architects or engineers (structural, MEP) with energy modeling experience
- A landscape architect with a specialty in native plant material
- A commissioning expert (if LEED employed)
- An engineer/architect with building modeling experience

The design team for a sustainably designed project tends to have a larger pool of talent than a typical architectural project. Wetlands scientists, energy efficient lighting consultants, native plant experts, or commissioning engineers are examples of the additional talent that may be added to sustainable design projects.

As with any architectural design, there is a hierarchy of design goals:

- *Initial imperatives* such as budget, timing, image, and program necessities
- *Subjective goals* such as a functionally improved and more pleasing work environment, pleasing color schemes, and landscaping that complements the architecture
- *Specific goals* such as more open space, more natural light, less water usage, and adjacency to public transportation

And with the inclusion of sustainability there may be additional goals:

- *Initiatives that are specific to sustainability* such as fewer toxins brought into the space, daylighting in all spaces with people occupancies, less overall energy consumed, less water usage, adjacency to public transportation, and improved indoor air quality

■ Desire to exceed existing standards such as ASHRAE, USGBC, or American Planning Association (APA)

RESEARCH AND EDUCATION

Is additional education and research necessary for a sustainable project?

Yes. Innovative HVAC systems, durable yet nontoxic materials, recycled materials, recyclable materials, native plant material, energy efficient lighting, and controls are examples of design components that are not normally designed and installed by general contractors and architectural consultants on typical projects.

Education of the Client

Sustainable design requires a new way of examining the architectural design process. Concepts such as life-cycle costing, recycled versus recyclable materials, non-VOC (volatile organic compounds) substances, daylighting, and alternate energy sources are among the several new concepts that the architect should discuss with the client before the design process commences.

It is critical that the client understands the sustainable process and is sympathetic to its potential economic and environmental benefits.

Education of the Project Team

Once the project has been assigned to an architect, but before the design process begins, the project team (architect, engineer, contractor, consultants, and owner) should assemble and discuss the project scope and objectives with all the project team members.

Establishing Project Goals

Among the many items included in the scope of work (including the extent of work, program elements, budget, and schedule) are the objectives for sustainable design.

For example, the architect and owner might establish goals for several environmental areas such as:

■ X percent reduction of energy usage from the established norm (see "Benchmarking" later in this section)

■ Improved lighting (less energy used and more efficient dispersal of indirect light with less glare)

■ Nontoxic and low VOC paint and finishes

■ Increased recycled content in materials such as carpeting, gypsum wallboard, ceiling tiles, metal studs, and millwork

■ High-efficiency (Energy Star) appliances

■ Wood elements are all certified wood products

■ Daylighting in all work/occupied spaces

As the leader of the project team, it is the architect's responsibility to include sustainable goals with the rest of the project scope of work.

A detailed explanation of the benefits of these sustainable design elements to all of the project team will ensure that they fully understand the design potential and economic implications of these concepts.

Verify Extent of Work

Sustainable design involves a more comprehensive approach to pre-project planning.

The LEED certification process will require record keeping and verification of the source of materials—a process that is beyond the normal design and construction work. For purposes of selecting a contractor and consultants, the team should be briefed on these additional obligations.

For example, the demolition process (if LEED certified) will require verification that materials have been sorted and delivered to an approved recycling organization. By contrast, the normal demolition process does not require recycling or verification that each material is sorted by type.

Clearly establishing the extent and type of effort required for each member of the sustainable design team is critical. The extent and type of effort will affect each member's ability to participate and their fees for services and construction work.

Energy and Optimization Modeling

Building shape, orientation, fenestration location, roof color, envelope configuration, and HVAC system efficiency are some of the variables in sustainable design projects that can be fine tuned with DOE-2 (U.S. Department of Energy's building analysis program) and other computer energy modeling programs.

The "fine tuning" of a project's energy components is one of the elements in the architect's design matrix that affects the final appearance, cost, and performance of the final design.

Energy modeling will not govern the final design. Issues such as compatible scale, color, texture, and functionality are still part of the architect's palette. But energy modeling is one additional factor that the architect will employ as part of the "best practices" approach to architecture.

In addition, modeling can assist in the cost analysis of a project. The fact that the modeling program is interactive helps the architect simultaneously adjust design elements to demonstrate alternate energy efficient solutions.

For example, energy modeling might allow the architect to demonstrate to the team that a more durable, aesthetically pleasing, and energy efficient building skin could be economically justified by reducing the size and cost of the mechanical system.

The ability to visually and numerically quantify the efficacy of trading certain design elements may be an effective tool for the architect when discussing the building design with the consultants and owner.

The Bid and Specification Process

The requirements of a sustainable design will often vary from a normal project.

For example, the millwork section of bid documents will normally specify the finish material, configuration of the design and methods of attachment, delivery, and installation. But the requirement of non-VOC glues and non-VOC substrate may confuse a potential bidder and cause that bidder to increase the bid price unnecessarily.

To facilitate the bidding and construction process, the architect should include the following:

- Simple definitions of sustainable elements—for example, what "VOC," "certified" wood product, or "daylighting" mean

- Explanations of specific characteristics of sustainable elements—for example, specifically state the standard that must be met (for example, Green Label Testing Program Limits, carpet's total VOC limit, that is, formaldehyde 0.05 (mg/m2)

- References of specific regulatory agency's information (name, address, e-mail, phone, and so on)—for example, the Carpet and Rug Institute, *www.carpet-rug.com*, 800-555-8846

- Examples of suppliers that could meet the sustainable standards indicated—in the case of sustainable products, there are at least two approaches to a list of suppliers for products:

  - Limit the installer to three to five suppliers of a product that is known to satisfy the sustainable design specifications.

    This approach assures the architect that the product will meet specified standards.

    (Note, however, that with the constantly changing nature of the emerging sustainable design market, a limited list could limit competition and the diversity of creative alternatives.)

  - Identify a list of qualified suppliers, but permit the bidder/contractor to submit alternative suppliers who satisfied the sustainable design criteria. This approach creates a more competitive environment, but it will require more effort of the architect to properly review and qualify the bids.

Changes and Substitutions

Every project is faced with the reality of time and budgetary pressures. And, in those instances, there may be situations when one product or design element may not be available in the form originally specified.

Sustainable designed projects require more stringent architectural supervision to ensure that original design standards are met.

For example, in the rush to project completion, the installer may claim that paints used for "touch up" of damaged areas are so small

that they may be installed with normal, higher-VOC paints. This minor transgression might jeopardize the integrity of the project and the ability to receive certification for LEED credits in certain areas.

ENERGY EVALUATION

In the climates of North America, buildings need some form of purchased energy (electricity, natural gas, oil) in order to operate. The architect works with a team to design strategies that may reduce the amount of purchased energy, reduce operating costs, and reduce the nation's dependence on imported fossil fuels.

The following are some design strategies that the sustainable design approach might employ to improve a building's energy performance. These elements are listed and briefly described.

Solar Design

Solar design is the age-old system of using sunlight or solar radiation to supply a portion of the building's heat energy. By a combination of techniques such as window and skylight design, location of internal thermal mass, and internal organization of the building's functions, solar design may replace some of the fossil fuel needed for heating and cooling buildings.

Passive solar systems is a category of solar design. Passive solar systems are those systems that permit solar radiation to fall on areas of the building that benefit from the seasonal energy conditions of the structure.

For example, some North American buildings are designed to reduce solar radiation gains from sunlight in the summer. Passive solar design relies on inherent qualities of the building's fenestration, massing, and orientation to capture sunlight.

Passive solar systems are usually categorized into direct or indirect gain systems.

Direct gain systems, as the title implies, are those systems that allow solar radiation to flow directly into the space needing heat.

A process commonly known as the "greenhouse effect" allows much of the sunlight that passes through the glass of the fenestration to be retained in the material it strikes (stone, concrete, wood, etc.) inside the building. Thus, south facing windows allow solar radiation to be directly gained and used inside the building.

Indirect gain systems operate when the sunlight first strikes a thermal mass that is located between the sun and the space. The sunlight absorbed by the mass is converted to thermal energy (heat) and then transferred into the living space.

There are basically two types of indirect gain systems: thermal storage walls and roof ponds. The difference is essentially the location—roof verses wall materials.

Passive solar design might employ several architectural strategies to facilitate the design:

- *Architectural sun control devices.* Overhangs or shading devices that have been designed to permit winter solar radiation from entering the building interior while blocking the higher angled, summer solar radiation from entering the building. Deciduous trees often perform the same function of permitting winter sunlight to enter and blocking much of the summer solar radiation with branches and leaves. Other examples include shutters, vertical projections or fins, awnings, trellises (especially with shading vegetation); and sunscreens (some with PV panels that both gather sunlight to convert into electricity and shade unwanted radiation from interior space in the warm months).

- *Light-colored roof systems.* Light-colored roofing materials reflect sunlight and reduce the amount of radiation that is absorbed through the roof into the interior space. Colors with higher reflectance (albedo) factors are preferred. For example, some cities in the United States require roof materials to have a minimum albedo rating of 0.65 (65 percent of the solar radiation is reflected back into the atmosphere). The urban heat island effect, caused by roofs, roads, and parking areas that absorb solar radiation during the day and retransmit the stored heat during the afternoon and evening, can be modified with light-colored roof systems.

  By designing these surfaces with light-colored and reflective material, the amount of heat energy stored in these materials is diminished and the urban heat island effect is reduced. Grass or vegetated roof areas have good insulating value and may also reduce the urban heat island effect and provide cooling through evapotranspiration.

- *Optimized building glazing systems.* Orientation, light transmittance factors, and U-value are all factors architects consider in selecting glazing. Glass that is low-E (emissivity) is desirable because it is coated with a material that allows a maximum amount of sunlight to be transferred through the glass and not reflected back into the atmosphere.

Lighting

The illumination of the interior of a sustainably designed building requires a holistic approach that balances the use of artificial and natural lighting sources.

Daylighting

Properly filtered and controlled solar radiation may provide a valuable source of illumination to a building interior. This process is called "daylighting" (simply having properly designed fenestration that allows natural sunlight to replace or dramatically reduce the need for artificial lighting).

Because unwanted sunlight (particularly in summer months) can also add to the internal heat load of a building, the architect must be careful to allow only beneficial sunlight and reduce unwanted solar heat gain. There are several techniques for controlling daylighting:

- Overhangs, fins, and other architectural shading devices

- Sawtooth (not bubble) skylight design, which allows the glass to face north for illumination, not south for solar heat gain

- Interior window shading devices, which allow solar gain during cool months, and the blocking of solar radiation during the warmer seasons

- Light shelves, which permit the daylight to reflect off the ceiling and penetrate farther into the interior without affecting views outside

Higher Efficiency Light Fixtures

In addition to a daylighting strategy, light fixtures that are more efficiently designed reduce energy cost and increase comfort, such as the following:

- Fixtures that use fluorescent or HID lamps, which provide more illumination per watt than incandescent lighting

- Fixtures that are designed to diffuse or bounce the illumination off the ceilings or internal reflectors, which are more efficient, cause less glare (especially in an environment with computer monitors), and save operating costs

- Fixtures that have higher efficiency (T-8) fluorescent bulbs, which produce more lumens per watt and thereby diminish the heat generated by lighting

- Fixtures that offer dimming or multiple switching capability, which permit the architect a more energy-efficient lighting design. Dimming or multiple switching fixtures allow the architect to design lighting patterns that blend nicely with daylighting opportunities. For example, an office with perimeter fenestration allows daylighting supplemented with overhead lighting that can be dimmed or reduced. The interior spaces, which are too far from the perimeter for daylighting, may be controlled with switches or dimmers that allow relatively higher levels of illumination. The result is an even illumination pattern, which saves on artificial lighting costs, by relying on daylighting at the perimeter.

- Fixtures that use higher efficiency lamps such as fluorescent, high intensity discharge (HID) sulfur lighting (exterior only)

- Fluorescent fixtures that use high efficiency electronic ballasts

Additionally, the architect may avoid less efficient incandescent lighting where possible; install task lighting to supplement diffused ambient lighting, and install LED (light emitting diode) lighting for exit signs. LED lighting lasts longer than incandescent and is far less expensive to operate.

Lighting Sensors and Monitors

Where possible, lighting costs can be diminished by installing light monitors that sense occupancy conditions. As long as the room contains people, the lights will remain on. If people leave, the sensor will wait for a few minutes, then shut off all the lighting in the room.

Lighting sensors can be designed to operate with a preference for motion, heat (from people), or desired time of occupancy.

Lighting Models

Computer lighting models are one option that allows the architect to simulate the levels of sunlight that penetrate into a building design, depending on the building location, varying times of year, fenestration orientation, and design.

By incrementally altering fenestration (skylights, windows, or light transport systems) and the artificial lighting system, the architect may optimize the daylighting and artificial lighting systems for the building.

Benchmarking

The U.S. Department of Energy provides "benchmark" information of total energy consumption in BTUs/SF for various kinds of buildings in the United States. These standards, or benchmarks, can be useful in the measuring of energy efficiency standards for various types of buildings:

For example:

- Average for all office buildings
  (pre-1990)   104.2
- Average for all office buildings
  (1990–1992)   87.4
- Average for all educational buildings
  (pre-1900)   87.2
- Average for all educational buildings
  (1990–1992)   57.1
- Average for all laboratory buildings
  (pre-1990)   319.2
- Average for all health care buildings (pre-1990)   218.5

Source: U.S. Department of Energy, Commercial Building Energy Consumption and Expenditures.

Benchmarking is a good way to alert the design team to the base energy standards for their design. It's a good place to start and ultimately a standard to beat. And, one can see from some of the comparisons (office and educational buildings), that some energy efficiency is occurring.

COMMISSIONING

Commissioning is an organized process to ensure that all building systems perform interactively according to the intent of the architectural and engineering design, and the owner's operating needs.

Commissioning usually includes all HVAC and MEP systems, controls, ductworks and pipe insulation, renewable and alternate technologies, life safety systems, lighting controls and daylighting systems, and any thermal storage systems. Commissioning also verifies the proper operation of architectural elements such as the building envelope, vapor and infiltration control, and gaskets and sealant used to control water infiltration.

Commissioning is a process required for LEED certification, but is a recommended procedure for any building involved with sustainable design procedures.*

INNOVATIVE TECHNOLOGIES

Besides the aforementioned issues of solar design, improved lighting systems, improved HVAC systems, and improved building massing and envelope design, there are several "innova-

---

* *Source: Commissioning Requirements for LEED Green Building Rating, Version 8. February 5. 1999; Sandra Mendler and William Odell,* The HOK Guidebook to Sustainable Design, *New York, John Wiley & Sons, Inc.: 2000, p. 71.*

tive technologies" that the architect can offer to the project team for consideration.

Groundwater Aquifer Cooling and Heating (AETS)

One alternative to full air-conditioning with chillers, which make heavy demands on electricity, is the aquifer thermal energy storage, which uses the differential thermal energy in water from an underground well to cool a building during summer and heat a building in the winter.

This is an efficient, relatively low-cost system, but it may require approval from the local environmental authority before installation.

Geothermal Energy

Where appropriate, heat contained within the earth's surface causes macrogeological events (such as underground geothermal springs or lava formations) that may be tapped to produce heat for adjacent structures.

In select locations, this heat energy can be transferred and conveyed to supplement a building's heating demand.

Wind Turbines

Small-scale wind machines used to generate electricity can be mounted on buildings or in open space nearby. These systems share several advantages:

- Relatively cost-effective
- Tested and established technology
- Systematic started-up
- Relatively high output

These systems share several disadvantages:

- Need a relatively high mast
- Require substantial structural support
- Present potential for noise pollution
- Visually intrusive

Photovoltaic (PV) Systems

The basis of the PV systems is the concept that electricity is produced from solar energy when photons or particles of light are absorbed by semiconductors.

Most PV systems are mounted to the building (either on the roof or as shading devices above fenestration). Currently, PV systems are not cost effective. But with promised government subsidy necessary to achieve an economy of scale, PVs may be a viable method of electrical production in the United States, Japan, and Germany in the near future.

Fuel Cells

Even though Sir William Grove invented the technology for the fuel cell in 1839, it has only recently been recognized as a potential power source for the future. The fuel cell claims to be the bridge between the hydrocarbon economy and the hydrogen-based society.

Fuel cells are electrochemical devices that generate direct current (DC) electricity similar to batteries. But, unlike batteries, they require a continual input of hydrogen-rich fuel. In essence, fuel cells are reactors that combine hydrogen and oxygen to produce electricity, heat, and water. They are clean, quiet, and emit no pollution when fed directly with hydrogen.

At the moment, fuel cell technology is still not cost effective for the commercial building market. Still, there seems to be a general feeling that hydrogen-based energy reactors will soon be an optional energy source.

Biogas

Biogas is produced through a process that converts biomass, such as rapid-rotation crops and selected farm and animal waste, to a gas that can fuel a gas turbine. This conversion process occurs through anaerobic digestion—the conversion of biomass to gas by organisms (like bacteria) in an oxygen-free environment.

Biogas has several advantages: it has relatively high energy production, it lends itself to both heat and power production, it creates almost zero carbon dioxide emissions, it virtually eliminates noxious odors and methane emissions, and it protects groundwater and reduces the landfill burden.

Small-Scale Hydro

Harnessing the energy from moving water is one of the oldest energy production systems in the world. In some locations, small-scale hydro power is a efficient and clean source of energy and is devoid of environmental penalties associated with large scale hydro projects. It allows small scale, local energy production, with relatively low cost.

Ice Storage Cooling Systems

One of the problems for energy supply companies is that the highest demand for electricity often coincides with the highest cooling demand.

The utilities would prefer to "flatten the curve" (to even out or flatten the measure of average daily energy demand). The fewer the number of peaks (high points of energy demand), the less the utilities have to bolster their power supply with expensive, supplemental fuels.

One way to reduce this peaking problem is to supplement a building's cooling capacity with an ice storage system.

An ice storage system has three components: a tank with liquid storage balls, a heat exchanger, and a compressor for cooling. The essence of the ice storage system is that the chilling and freezing of the ice balls occurs at night (when the cost of energy is lower due to lower demand). During the day, the cool temperatures, stored in the ice, are transmitted into the building's cooling system.*

CONCLUSION

The knowledge of environmental systems has become essential to the architect's design palette. Buildings that take advantage of natural systems such as sun, wind, rain, groundwater, topography, and climate are more elegant solutions. Architectural designs that incorporate natural systems, in conjunction with contemporary technologies, are in the tradition of architects providing spatial solutions with the most innovative contemporary thinking available.

Buildings with this approach operate more efficiently, integrate effectively into their local environment, and tend to produce spaces that are more pleasing to work and live. Knowledge of integrated or holistic design principals is not a limitation but another set of tools to produce humane, efficient, healthy, and aesthetically compelling architecture.

---

* Source: Peter F. Smith, *Sustainability at the Cutting Edge*, Jordan Hill, Oxford: Architectural Press, an Imprint of Elsevier Science, 2003.

LESSON 9 QUIZ

1. Sustainable design is primarily concerned with which of the following issues?

   I.   Economics
   II. Aesthetics
   III. Environment
   IV. Mechanical systems

   A. III
   B. I, II, and III
   C. I and III
   D. All of the above

2. *The Natural Step* is an approach to the environment that follows which of the following principles?

   I.   The biosphere affecting humans is a relatively stable and resilient zone that includes five miles into the earth's crust and five miles into the troposphere.
   II. Improved technologies have dramatically increased the number and quantity of available natural resources.
   III. Toxic substances released into either the sea or atmosphere will only influence areas adjacent to the toxic source.
   IV. Using building materials that are recycled is an adequate sustainable design approach.

   A. I
   B. II
   C. II and IV
   D. None of the above

3. The planning phase of a sustainably designed architectural project should include which of the following elements?

   I.   Native landscaping that is aesthetically pleasing and functional
   II. Designing structures in the floodplain that can resist the forces of flood waters
   III. Consideration of sun orientation, topographic relief, and the scale of adjacent buildings
   IV. Locating projects within existing neighborhoods that are adjacent to public transportation

   A. I and II
   B. I and III
   C. I, III, and IV
   D. All of the above

4. The Ahwahnee principles include which of the following ideas?

   I.   Communities with only residential use should be relegated to areas outside the central business district.
   II. Preserved open spaces should be either wildlife habitats or recreational areas.
   III. Transportation planning should include roads, pedestrian paths, bike paths, and mass transit systems.
   IV. Job creation and economic diversity is a desired goal.

   A. I
   B. II, III, and IV
   C. III and IV
   D. None of the above

5. Life cycle costing is an economic evaluation of architectural elements that includes which of the following factors?

   I. First cost

   II. Maintenance and operational costs

   III. Repair costs

   IV. Replacement cost

   A. I

   B. II, III, and IV

   C. II and IV

   D. All of the above

6. LEED is concerned with which of the following? Check all that apply.

   A. Indoor air quality

   B. Storm water

   C. Construction costs

   D. Tax benefits

   E. Innovative energy systems

   F. Aesthetic design

7. Which of the following is a consultant who might be employed in a sustainable design project?

   I. Wetlands engineer

   II. Energy commissioner

   III. Landscape architect

   IV. Energy modeling engineer

   A. I

   B. I and II

   C. II, III, and IV

   D. All of the above

8. Sustainable design may require research and education that is beyond a normal architectural project. Which of the following is part of this process?

   I. Energy modeling

   II. Education of the client

   III. Art selection

   IV. Selection of energy efficient appliances

   A. I and IV

   B. I and II

   C. I, II, and IV

   D. All of the above

9. Sensitivity to the nuances of site conditions is key to sustainable design. Which of the following are site conditions the architect should examine in the design process?

   I. Solar orientation

   II. Decorative landscaping

   III. Scale and style of adjacent structures

   IV. Groundwater conditions

   A. I and II

   B. I, III, and IV

   C. I and III

   D. All of the above

10. Sustainably designed architecture requires attention to which of the following building elements?

    I. Solar shading devices

    II. Urban heat island effect

    III. Increased parking

    IV. Fenestration and glazing

    A. I, II, and IV

    B. I and IV

    C. I and II

    D. All of the above

Part II

The Graphic Vignettes

THE NCARB SOFTWARE

INTRODUCTION

There is a wide variety of programs used by candidates at the firms in which they work. Therefore, an essential part of every candidate's preparation is to practice using the examination's computer tools. Candidates can download this software from the NCARB Web site (*www.ncarb.org*). This program contains tutorials and sample vignettes for all the graphic portions. Spend all the time necessary to become familiar with this material in order to develop the necessary technique and confidence. You must become thoroughly familiar with the software.

The drafting program for the graphic portions is by no means a sophisticated program. While this may frustrate candidates accustomed to advanced CAD software, it is important to recall that NCARB aimed to create an adequate drafting program that virtually anyone can use, even those with no CAD background at all.

VIGNETTE SCREEN LAYOUT

Each vignette has a number of sections and screens with which the candidate must become familiar. The first screen that appears when the vignette is opened is called the Vignette Index and starts with the Task Information Screen. Listed on this screen are all the components particular to this vignette. Each component opens a new screen when the candidate clicks on it with the mouse. A menu button appears in the upper left corner of any of these screens that returns you to the Index Screen. Also available from the Index Screen is a screen that opens the General Test Directions Screen, which gives the candidate an overview of the procedures for doing the vignettes. Here are the various screens found on the Index Screen:

- **Vignette Directions** (found on all vignettes)—describes the procedure for solving the problem

- **Program** (found on all vignettes)—describes the problem to be solved

- **Tips** (found on all vignettes)—gives advice for approaching the problem and hints about the most useful drafting tools

- **Code**—gives applicable code information if required by the vignette

- **Sections**—typically found on the Stair Design Vignette and shows a section through the space in which the stair will be located

- **Lighting Diagrams**—found on the Mechanical and Electrical Plan vignette to show light fixture distribution patterns

The beginning of each vignette lesson in this study guide provides a more detailed description of each vignette screen.

To access the actual vignette problem, press the space bar. This screen displays the problem and all the computer tools required to solve it. Toggle back and forth between the Vignette Screen and one of the screens from the Index Screen at any time by simply pressing the space bar. This is not as convenient as viewing both the drawing and, say, the printed program adjacent to each other at the same time. Thus it is a procedure that the candidate must become familiar with through practice. Also, some vignettes are too large to be displayed all at once on the screen. In this case use the scroll bars to move the screen up and down or left and right as needed. The Zoom Tool is also helpful.

THE COMPUTER TOOLS

There are two categories of computer tools found in the ARE graphic portions:

- Common Tools
- Tools specific to each vignette

The Common Tools, as the name implies, are generally present in all the tests and allow a candidate to draw lines, circles, and rectangles, adjust or move shapes, undo or erase a previously drawn object, and zoom to enlarge objects on the screen. There is also an on-screen calculator and a tool that lets you to erase an entire solution and begin again.

Vignette-specific tools include additional tools that enable the candidate to turn on and off layers, rotate objects, and set elevations or roof slopes. In addition to these extra tools, each vignette also includes specific items under the draw tool required for the vignette, such as joists or skylights. Become an expert in the use of each tool.

Each tool is dependent on the mouse; there are no "shortcut" keys on the keyboard. Press the computer tool first to activate it, then select the item or items on the drawing to be affected by the tool, and then re-click the computer tool to finish the operation. Spend as much time as required to become completely familiar with this drafting program. The Common Tools section of the practice vignettes available from NCARB is particularly useful for helping you become familiar with the computer tools. Three things to note: the left mouse button activates all tools; there is no zoom wheel on the mouse, nor an associated tool on the program; and the shift key activates the Ortho Tool.

The standard computer tools and their functions are shown in Figure 10.1.

BRINGS UP A MENU OF ITEMS, SUCH AS
ROOMS, COLUMNS, DOORS, ETC.; SOME
MENU ITEMS, SUCH AS *JOISTS,* MAY LEAD
TO SUBMENU ITEMS, SUCH AS *SPACING*

CHANGES THE SIZE OR SHAPE OF
A SINGLE OBJECT, OR RELOCATES
A PREVIOUSLY DRAWN OBJECT

MOVES A SELECTED ARRANGEMENT
OF OBJECTS AS A GROUP

ROTATES PREVIOUSLY DRAWN OBJECTS

BRINGS UP A SUBMENU, ALLOWS ACCESS TO
OTHER LAYERS OR OTHER FLOOR PLANS, ALLOWS
MULTIPLE LAYERS TO BE VIEWED AT THE SAME
TIME OR JUST ONE LAYER TO BE ISOLATED

VERIFIES SEVERAL CRITICAL CONDITIONS,
SUCH AS OVERLAPPING SPACES

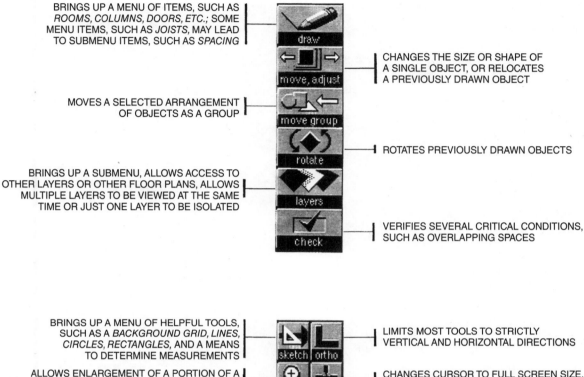

BRINGS UP A MENU OF HELPFUL TOOLS,
SUCH AS A *BACKGROUND GRID, LINES,
CIRCLES, RECTANGLES,* AND A MEANS
TO DETERMINE MEASUREMENTS

LIMITS MOST TOOLS TO STRICTLY
VERTICAL AND HORIZONTAL DIRECTIONS

ALLOWS ENLARGEMENT OF A PORTION OF A
DRAWING TO PRODUCE DETAILED WORK; EMPLOYS
A PICK BOX TO ENLARGE ONLY WHAT IS SELECTED AND
THE OPTION TO RETURN TO THE ORIGINAL VIEW BY
CLICKING ON THE TOOL A SECOND TIME

CHANGES CURSOR TO FULL SCREEN SIZE,
WHICH HELPS ALIGN OBJECTS

DELETES THE LAST OPERATION COMPLETED

REMOVES PREVIOUSLY DRAWN OBJECTS

PROVIDES INFORMATION FOR SELECTED
OBJECTS, SUCH AS *SIZE, AREA, ANGLE, ETC.*

BRINGS UP A SCIENTIFIC CALCULATOR;
CLICK TO DISPLAY, CLICK AGAIN TO HIDE

RETURNS TO THE PROGRAM SCREEN–
SERVES THE SAME FUNCTION AS THE SPACE
BAR ON THE COMPUTER KEY BOARD

ERASES ALL DRAWING FROM THE SCREEN TO
BEGIN A NEW SOLUTION, IT IS SUGGESTED THAT
THIS TOOL ONLY BE USED AS A LAST RESORT

SELECTS ANOTHER VIGNETTE IN THE SAME
SECTION OR ALLOWS AN EXIT FROM THE EXAM

Figure 10.1

TAKING THE EXAM

INTRODUCTION

Preparation for the ARE usually begins several months before taking the actual exam. The first step is to submit an application for registration with your state board or Canadian provincial association. Most, but not all, registration boards require a professional degree in architecture and completion of the Intern-Development Program (IDP) before a candidate is allowed to begin the exam process. Because the processing of educational transcripts and employment verifications may take several weeks, begin this process early. The registration board will review a candidate's application to determine whether the candidate meets the eligibility requirements.

SCHEDULING THE EXAM

The exams are available to eligible candidates at virtually any time, because test centers are open nearly every day throughout the year. However, it is the responsibility of the candidate to contact a test center to schedule an appointment. This must be done at least three days prior to the desired appointment time, but it is probably more sensible to make an appointment a month or more in advance. It is not necessary to take the test in the same jurisdiction in which you intend to be registered. Someone in San Francisco, for example, could conceivably combine test-taking with a family visit in Philadelphia.

FINAL PREPARATION

Candidates are advised to complete all preparations the day before their appointment in order to be as relaxed as possible before the upcoming test. Avoid last-minute cramming, which in most cases does more harm than good. The graphic portions not only test design competence, but also physical and emotional endurance. You must be totally prepared for the strenuous day ahead, and that requires plenty of rest and composure.

One of the principal ingredients for success on this exam is confidence. If you have prepared in a reasonable and realistic way, and if you have devoted the necessary time to practice, you should approach the Building Design & Construction Systems division with confidence.

EXAM DAY

Woody Allen once said that a large part of being successful was just showing up. That is certainly true of the licensing examination, where you must not only show up, but also be on time. Get an early start on exam day and arrive at the test center at least 30 minutes before the scheduled test time. Getting an early start enables you to remain in control and maintain a sense of confidence, while arriving late creates unnecessary anxiety. If you arrive 30 minutes late, you may lose your appointment and forfeit the testing fee. Most candidates will begin their test session within one-half hour of the appointment time. You will be asked to provide a picture identification with signature and a second form of identification. For security reasons, you may also have your picture taken.

THE EXAM ROOM

Candidates are not permitted to bring anything with them into the exam room: no reference materials, no scratch paper, no drawing equipment, no food or drink, no extra sweater, no cell phones, no digital watches. You are permitted to use the restroom or retrieve a sweater from a small locker provided outside the exam room. Each testing center will have its own procedure to follow for such needs. A calcula-

tor is provided as part of the drafting program. Scratch paper will be provided by the testing center. The candidate might wish to request graph paper, if available.

Once the candidate is seated at an assigned workstation and the test begins, the candidate must remain seated, except when authorized to leave by test center staff. When the first set of vignettes is completed, or if time runs out, there is a mandatory break, during which you must leave the exam room. Photo identification will be required when you reenter the exam room for the next set of vignettes. At the conclusion of the test, staff members will collect all used scratch paper.

Exam room conditions vary considerably. Some rooms have comfortable seats, adequate lighting and ventilation, error-free computers, and a minimum of distractions. The conditions of other rooms, however, leave much to be desired. Unfortunately, there is little a candidate can do about this, unless, of course, a computer malfunctions. Staff members will try to rectify any problem within their control.

EXAM ROOM CONDUCT

NCARB has provided a lengthy list of offensive activities that may result in dismissal from a test center. Most candidates need not be concerned about these, but for those who may have entertained any of these fantasies, such conduct includes:

- Giving or receiving help on the test
- Using prohibited aids, such as reference material
- Failing to follow instructions of the test administrator
- Creating a disturbance

- Removing notes or scratch paper from the exam room
- Tampering with a computer
- Taking the test for someone else

BEGIN WITH THE PROGRAM

You can either solve the vignettes in the order they are presented or build confidence by starting with one that looks easier to you. Only you know what works best for you; the practice software should give you a sense of your preferred approach.

Every vignette solution begins with the program. Read the entire program carefully and completely, and consider every requirement. During this review, identify the requirements, restrictions, limitations, code demands, and other critical clues that will influence the solution. Feel free to use scratch paper to jot down key points, data, and requirements. This will help ensure that you understand and meet all the requirements as you develop your solution.

Every vignette problem has two components: the written program and a graphic base plan. Both components are complementary and equally important; together, they completely define the problem. Candidates should not rush through a review of the program and base drawing in an attempt to begin the design sooner. It is more important to understand every constraint and to be certain that you have not overlooked any significant detail. Until you completely understand the vignette, it is pointless to continue.

GENERAL STRATEGIES

The approach to all vignette solutions is similar: Work quickly and efficiently to produce a solution that satisfies every programmatic requirement. The most important requirements are those that involve compliance with the code, such as life safety, egress, and barrier-free access.

Another important matter is design quality. Strive for an adequate solution that merely solves the problem. Exceptional solutions are not expected, nor are they necessary. You can only pass or fail this test, not win a gold medal. Produce a workable, error-free solution that is good enough to pass.

During the test session, candidates will frequently return to the program to verify element sizes, relationships, and specific restrictions. Always confirm program requirements before completing the vignette, so that oversights or omissions may be corrected while there is still time to do so. Candidates must always keep in mind the immutability of the program. That is, never—under any circumstances—modify, deviate from, or add anything to the program. Never try to *improve* the program. Lastly, taking time at the end of each section to review all the vignettes can help to eliminate small errors or omissions that could tip the balance between a passing and a failing grade.

Candidates should have little trouble understanding a vignette's intent. However, the true meaning of certain details may be ambiguous and open to interpretation. Simply make a reasonable assumption and proceed with the solution.

While candidates will necessarily employ their own strategies for ordering the vignettes, what follows are some ideas others have found helpful.

1. Start with the vignette you feel most capable of completely quickly and competently. This will boost your confidence for the remaining vignettes.

2. Try to solve each problem in 10 minutes less than the allotted time. Use this additional time to review your solution.

3. Create a set of notes or a chart for each problem.

THE TIME SCHEDULE

The most critical problem on the exam is *time*, and you must use that fact as the organizing element around which any strategy is based. The use of a schedule is essential. During the preparation period, and especially after taking a mock exam, note the approximate amount of time that should be spent on each vignette solution. This information must then become a candidate's performance guide, and by following it faithfully, the candidate will automatically establish priorities regarding how your time will be spent.

It is important to complete each vignette in approximately the time allotted. You cannot afford to dwell on a minor detail of one vignette while completely ignoring another vignette. Forget the details; do not strive for perfection, and be absolutely certain to finish the test. Even the smallest attempt at solving a vignette will add points to your total score.

Vignettes have been designed so that a reasonable solution for each of the problems can be achieved in approximately the amount of time shown in the *ARE Guidelines*. These time limits are estimates made by those who created this test. In any event, a 45-minute-long vignette may not necessarily take 45 minutes to complete. Some can be completed in 30 minutes, while others may take an hour or longer. The time required depends on the complexity of the problem and your familiarity with the subject matter. Some candidates are more familiar with certain problem types than others, and because candidates' training, experience, and ability vary considerably, adjustments may have to be made to suit individual needs. It should also be noted that within each exam section, the time allotted for two vignettes may be used at the candidate's discretion. For example, in a three-vignette section that allots 150 minutes, NCARB recommends spending one hour on one vignette and 45 minutes on the other two. However, you may actually spend 80 minutes on one problem and 35 minutes on the other two.

Candidates who are aware of the time limit are more able to concentrate on the tasks to be performed and the sequence in which they take place. You will also be able to recognize when to begin the next vignette. When the schedule tells you to stop working on one vignette and move on to the next, you will do so, regardless of the unresolved problems that may remain. You may submit an imperfect solution, but you *will* complete the test.

Only solve what the program asks you to solve, and don't use real-world knowledge, such as specific building code requirements.

TIME SCHEDULE PROBLEMS

It is always possible that a candidate will be unable to complete a certain vignette in the time allotted. What to do in that event? First, avoid this kind of trouble by adhering to a rigid time schedule, regardless of problems that may arise. Submit a solution for every vignette, even if some solutions still have problems or are incomplete.

Candidates are generally able to develop some kind of workable solution in a relatively short time. If each decision is based on a valid assumption and relies on common sense, the major elements should be readily organized into an acceptable functional arrangement. It may not be perfect, and it will certainly not be refined, but it should be good enough to proceed to the next step.

MANAGING PROBLEMS

There are other serious problems that may arise, and while each is potentially fatal, they must be managed and resolved. Consider the following:

- The candidate has inadvertently omitted a major programmed element.
- The candidate has drawn a major element too large or too small.
- The candidate has ignored a critical adjacency or other relationship.

The corrective action for each of these issues will depend on the seriousness of the error and when the mistake is discovered. If there is time, one should rectify the design by returning to the point at which the error occurred and begin again from there. If it is late in the exam and time is running out, there may simply be insufficient time to correct the problem. In that case, continue on with the remainder of the exam and attempt to provide the most accurate solutions for the remaining vignettes. The best strategy, of course, is to avoid critical mistakes in the first place, and those who concentrate and work carefully will do so.

WORKING UNDER PRESSURE

The time limit of the graphic portions creates subjective, as well as real problems. This exam generates a unique psychological pressure that can be harmful to performance. While some designers thrive and do their best work under pressure, others become fearful or agitated under the same conditions. It is perfectly normal to be uneasy about this important event; and although anxiety may be a common reaction, it is still uncomfortable.

Candidates should be aware that pressure is not altogether a negative influence. It may actually heighten awareness and sharpen abilities. In addition, realize that, as important as this test may be, failure is not a career-ending event. Furthermore, failure is rarely an accurate measure of design ability; it simply means that you have not yet learned how to pass this difficult exam.

EXAMINATION ADVICE

Following is a short list of suggestions intended to help candidates develop their own strategies and priorities. We believe each item is important in achieving a passing score. The *ARE Guidelines,* available from the NCARB Web site, also lists suggestions for examination preparedness.

- **Get an early start.** Begin preparation early enough to develop confidence by the time you are scheduled to take the exam. Arrive at the exam site early and be ready to go when the test begins.
- **Complete all vignettes.** Incomplete solutions risk failure. Complete every problem, even if every detail is not complete or perfect.

- **Don't modify the program.** Never add, change, improve, or omit anything from a program statement. Never assume that there is an error in the program. Verify all requirements to ensure complete compliance with every element of the program. If ambiguities exist in the program, make a reasonable assumption and complete your solution.

- **Develop a reasonable solution.** Since most vignettes generally have one preferred solution, solve the problem in the most direct and reasonable way. Never search for a unique or unconventional solution, because on this exam, creativity is not rewarded.

- **Be aware of time.** The strict time constraint compels you to be a clock-watcher. Never lose sight of how much time you are spending on each vignette. When it is time to proceed to the next problem, quit and move on to the next vignette.

- **Remain calm.** This may be easier said than done, because this type of experience often creates stress in even the most self-assured candidate. Anxiety is generally related to fear of failure. However, if a candidate is well prepared, this fear may be unrealistic. Furthermore, even if the worst comes to pass and you must repeat a division, all it means is that your architectural license will be delayed for a short period of time.

ACCESSIBILITY/RAMP VIGNETTE

12

INTRODUCTION

The Accessibility/Ramp vignette deals with handicapped accessibility. This vignette tests a candidate's ability to apply conventional handicapped requirements to the design of a ramp and stair system. Candidates are given a program, code requirements, and a base plan on which they are required to develop a code-compliant ramp and stair system between two different building elevations. Your success depends on compliance with code requirements, appropriateness, and efficiency of the layout.

Keep in mind that the critical underlying life safety issue addressed in this vignette is to provide an accessible means of egress.

VIGNETTE INFORMATION

The Accessibility/Ramp vignette comprises a partial floor plan in which two portions of a building, or two different buildings, with different floor elevations, are to be connected by a new ramp and a separate stair. In addition, the problem usually requires you to locate a wall with an exit door at one level or the other. Begin with the index screen that offers access to the following screens:

■ **Vignette Directions**—These direct a candidate to complete the given floor plan by developing a ramp and stair system in accordance with the program information. Provide ramps, stairs, railings, walls, doors, and the elevations of all landings.

■ **Program**—This screen describes two buildings, or two parts of a building, with different elevations, for which a candidate is required to provide an accessible circulation system. Typically, you are not allowed to have the ramp or stair extend into the existing space.

■ **Code**—Unlike the first two vignettes, this vignette includes a code screen. The code screen contains the only code-related criteria you are permitted to use, and it includes the following categories:

Maneuvering Clearances—Clearances at both sides of a passage door are illustrated, as is the turning space required by a wheelchair.

Ramps—Allowable ramp width, slope, and landings.

Stairways—Allowable stair width, tread depth, and riser height.

Doors—Allowable door width and direction of swing.

Rails—Required location, continuity, extensions, and projections of both guardrails and handrails.

■ **Tips**—The number of risers must be calculated before drawing any stairs. When laying out a stairway, the tread depth is automatically displayed in the information area at the bottom of the screen. Candidates are also reminded that there are different floor elevations at various parts of the plan. Useful computer tools for this vignette include the *zoom tool, full-screen cursor,* and the *sketch* tools.

The base plan for this vignette is found on the **work screen**, which is accessed by clicking the space bar. The partial floor plan is a relatively simple drawing that contains four or five spaces and two different floor elevations. Contained under the draw tool are the specific elements required, such as walls, ramps, stairs, landings, railings, and doors.

DESIGN APPROACH

In the words of NCARB, the Accessibility/Ramp vignette is an easy problem to solve, because it simulates the kind of activity most architects deal with at work on a daily basis. While it might be one of the less difficult vignettes, it is by no means easy. Furthermore, not all architectural employees work with accessibility standards each day. Nevertheless, issues of accessibility are critical in designing usable spaces, and all candidates must be familiar with the legal standards of accessibility and know how to apply them in an architecturally appropriate way.

The objective of an accessibility problem is clearly stated in the program. The solution must comply with the code regulations found on the code screen. A candidate must competently apply all the pertinent rules, standards, and requirements.

DESIGNING A RAMP

Since an Accessibility/Ramp vignette includes both a ramp and a stairway, which element should be laid out first: the ramp or the stair? Actually, the first element you should lay out is a landing at either the high or low floor elevation, or more than one landing, if need be. In many cases, a single landing may serve both the ramp and the stair, which are generally near each other. Once the landing size is established, based on the allowable stair width and the existing floor elevation, then lay out the ramp, because it takes much more space than the stairway.

Ramps found in vignette problems may be within a building, such as a change in floor levels, or outside a building, such as between a sidewalk and a front porch. In either case, the principles and requirements are much the same. For example, you may be asked to provide both a ramp and stairs to get from the interior of one building to a higher adjacent structure, or to circulate from a paved alley to a 36-inch-high loading dock. Ramp problems always include the design of a stairway, which involves the same change in elevation.

The design of a ramp begins at whichever end of the ramp has a known or fixed location, such as a doorway or the end of a corridor. In some cases, the fixed locations of both ends may be known. Laying out a ramp involves several factors. First, you must know the vertical distance between the ends of the ramp, which is always given. Second, you must know the slope of the ramp. Generally, a ramp is defined as any accessible route with a slope greater than 1:20, but in no case may the slope exceed 1:12. In the case of Accessibility/Ramp vignettes, candidates should always use the ramp slope of 1:12, unless there is a persuasive reason to do otherwise.

Besides the vertical distance and slope, there are regulations governing ramps that must be considered, and these are included on the code screen. For example, the minimum clear width of a ramp may be given as 44 inches, although some ramps may be as narrow as 36 inches. Always begin a solution by using the minimum ramp width permitted by code. Another important restriction is that a ramp may not rise more than 30 inches before creating a level landing. At a slope of 1:12, this means that the maximum uninterrupted ramp length is 30 feet. The landing must be no less than five feet long, and if the ramp changes direction, the landing must be no less than five feet square. These dimensions allow a wheelchair to make a complete 180-degree turn.

Handrails are required on both sides of a ramp whenever the ramp's rise exceeds six inches. Handrails must be continuous and extend 12 inches beyond the top and bottom of the ramp. Handrail extensions are important, and candidates should be familiar with their application. For example, a handrail extension must never project into a circulation path or beyond a setback line. In some cases, it may wrap around a corner rather than create an improper or dangerous protrusion.

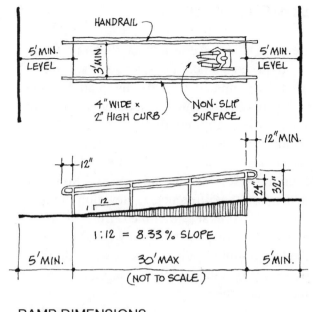

RAMP DIMENSIONS

Figure 12.1

Because ramps take a considerable amount of space, their layout requires common sense planning. For example, a ramp that sprawls over a large portion of the plan would be considered inefficient. In most cases, a switchback ramp takes the least amount of space and is also the least obtrusive. However, a number of short ramp lengths with an excessive number of switchbacks will cause circulation problems for the disabled. Also consider the space that remains after placement of the ramp: this left-over space should be usable for circulation or perhaps a well-proportioned planter or pool.

Once you have determined where the ramp begins, be certain there is a five-foot-deep level landing and proceed to lay out the ramp in a straight line. If the total rise exceeds 30 inches, end the ramp after running 30 feet, create a level landing five feet deep, and begin a new length of ramp. This new length of ramp will generally turn at 90° or 180°, in which case the level landing must be a minimum of five feet square.

STAIR LAYOUT

Because stairs take less space than ramps, fit the required stairs in the space that remains after the ramp is designed. The tread depth will generally be calculated automatically for you, but keep in mind that 11- or 12-inch deep treads are standard, with 11 inches being the minimum. The height of risers may be between four and seven inches, and your solution will depend on the difference in elevation in the problem. For example, if the difference in elevation were 5'-0" and you wished to create the least number of risers, you would divide 60 inches by the maximum riser height, seven inches, and the result would be 8.57 risers. In this case, you would design a stairway with nine risers (the next higher round number), and consequently, each riser would be $60 \div 9 = 6.67$ inches high. Always use the maximum allowable riser height to develop a solution with the fewest number of steps.

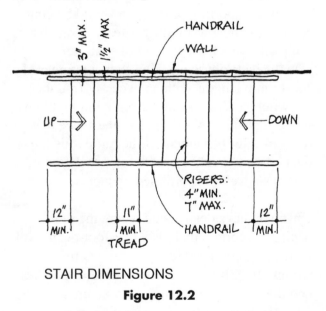

STAIR DIMENSIONS

Figure 12.2

The least dimension of a stairway landing must not be less than the required width of the stairs, which in most cases is 44 inches. Stairways also require handrails on both sides of the run, and these, too, must be continuous. If the stair-

way exceeds 88 inches in width, an intermediate handrail is required. Stair handrails must project beyond the risers at top and bottom by at least 12 inches. This projection may be greater, if specified by the program.

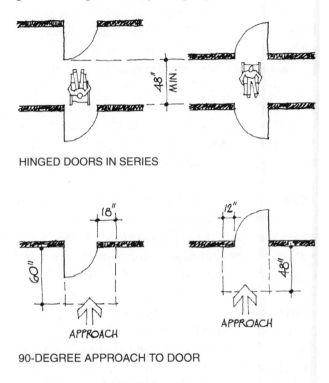

HINGED DOORS IN SERIES

90-DEGREE APPROACH TO DOOR

MANEUVERING CLEARANCES

Figure 12.3

DOOR CLEARANCES

Among the code-related data presented are the accessible maneuvering clearances at doors. The most common of these is hinged doors in series, which require 48 inches of clear space between two doors. There are also illustrations shown for inswinging and outswinging doors. In each situation, a clear space of a specific dimension is required for maneuvering. For example, a handicapped user directly approaching an outswinging door (that is, at 90° to a door swinging toward the approaching user) requires 18 inches of clear space at the latch side of the door.

Finally, certain elements are considered mandatory and, if lacking, the candidate's solution will be penalized. For example, a 5'-0" turning radius is required for the circulation of a wheelchair. The actual circle need not be shown, but you must indicate, throughout the entire circulation path, that sufficient space exists to turn a wheelchair in any direction or completely around. Another required element is guardrails, which must be shown on open sides of landings, floor surfaces, ramps, and stairways. Guardrails are similar to handrails in that they must be continuous and offer protection wherever hazardous changes in grade exist. Become familiar with these door clearances in order to feel confident in designing an appropriate door approach without depending on which maneuverability clearance diagrams will be provided on the Tips Screen in the vignette.

KEY COMPUTER TOOLS

- **Zoom Tool**—This tool will help the candidate set the ramp and stair components with greater accuracy.

- **Information Area**—At the bottom of the Vignette Screen is a space that indicates information about elements you are drawing or have highlighted.

- **Double Click**—Double-clicking the mouse button on overlapping elements on the screen will isolate the one desired.

VIGNETTE 1 ACCESSIBILITY/RAMP

Develop an accessible ramp and stair system to connect Building One and Building Two, as shown on the partial floor plan. Provide a new door between the Building One Lobby and the Access Area that will permit Building One occupants to circulate through the Access Area and exit from the Building Two Lobby. The Lobby elevation of Building One is 18 inches higher than the Lobby of Building Two, and construction to accommodate this change must be located in the Access Area. The completed plan should conform to all program and code requirements.

Program Requirements
General

- The minimum width of an exit route shall be not less than 44 inches.
- All required maneuvering clearances shall be on level surfaces.
- Indicate the elevations of all new landings.
- Wheelchairs require a minimum 60-inch diameter turning space.

Ramp

- The minimum ramp width shall be not less than 44 inches, less handrail projections.
- The maximum ramp slope shall be 1 vertical to 12 horizontal.
- Level landings at the top and bottom of each run shall be no less than the required width of the ramp, and no less than 60 inches long in the direction of travel.

Stairs

- The minimum stair width shall be not less than 44 inches, less handrail projections.
- The least dimension of landings shall be no less than the required stair width.
- All steps shall have a uniform tread depth of 11 inches minimum, and a uniform riser height of between 4 and 7 inches.

Handrails

- Handrails shall be provided on both sides of ramps and stairs.
- Handrails shall be continuous within each run.
- Noncontinuous handrails for ramps and stairs shall extend horizontally at least 12 inches beyond the top and bottom of each run.
- Open sides of landings shall be protected by a continuous rail.

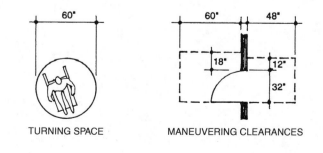

TURNING SPACE   MANEUVERING CLEARANCES

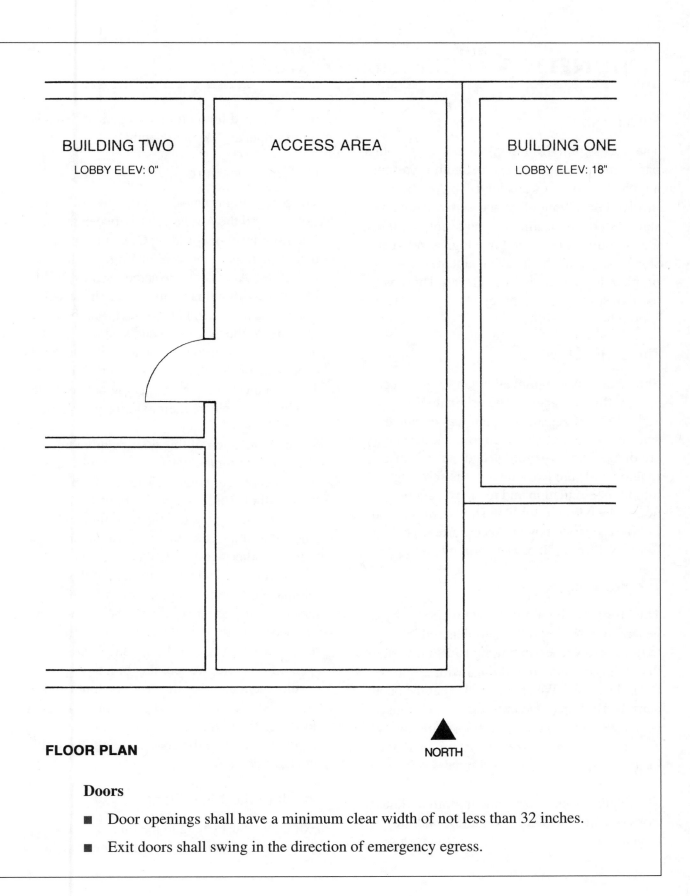

FLOOR PLAN

▲
NORTH

Doors

- Door openings shall have a minimum clear width of not less than 32 inches.

- Exit doors shall swing in the direction of emergency egress.

VIGNETTE 1   ACCESSIBILITY/RAMP

Introduction

The Accessibility/Ramp vignette comes from a paper-and-pencil mock exam, and it represents a typical exercise of this kind. The problem involved the layout of an accessible ramp and stair system connecting two adjacent buildings. Our solution is presented in a logical progression of steps in which each element of the problem is considered in sequence. Those who learn to solve vignette problems in this way should have few difficulties.

The Exam Sheet

Shown are the program and plan for the Accessibility/Ramp vignette. The original sheet was 12″ × 18″ in size, and the plan was drawn at a scale of ¼″ = 1′-0″. The program requirements are divided into categories, and beneath these requirements are two illustrations showing important circulation and maneuvering restrictions. The partial floor plan shows two building Lobbies separated by an Access Area. The two Lobbies differ in floor elevation by 18 inches.

The Program

The Program Requirements are preceded by a general description of the problem, which is to develop an accessible ramp and stair system that will connect two buildings with different floor elevations. We are also told that the occupants of Building One must circulate through the Access Area and exit through Building Two. The required ramp and stair system must be located within the prescribed Access Area.

The requirements specify the minimum dimensions of exit routes, ramps, stairs, handrail projections, and doors. You are also given recommended limits for stair risers and treads, ramp slope, and landing depths.

Design Procedure

The partial floor plan shows the two building Lobbies and the Access Area between them. The floor level of Building One is 18 inches higher than the floor level of Building Two, and the Access Area is on the same level as Building Two. Within this relatively small Access Area, which scales about 10 feet × 25 feet, we must fit the new ramp and stairway. First, however, in order to provide an exit path for the occupants of Building One through Building Two, we must cut a door opening between the two buildings where their exterior walls adjoin.

Since the location of this door opening will affect all subsequent design, we should carefully analyze our choices. The length of the west wall of Building One is about 17 feet, and, considering the Lobby of Building One, a new door may penetrate that wall at any point. In other words, whether a new door is placed at either end or the middle of the wall will have little effect on the Lobby itself. As far as the Access Area is concerned, however, a new door in the middle of its east wall would restrict development in an already restricted space. Therefore, we feel it is best to begin our development at either the north or south end of the space. Clearly, the best location for the new door is at the north end, because at that location the door will be in the corner of both the Lobby and the Access Area.

Locating the New Door

Exit doors must swing in the direction of emergency egress. Thus, the new door must swing out from the Lobby of Building One and into the Access Area.

We next return to the illustration in the program that shows the required maneuvering clearances. Although no clearance space is required at the hinge side of a door, we place the door an arbitrary 12 inches from the wall to allow for construction (jamb, trim, etc.). We use the allowable minimum width of 32 inches for the door and swing it against the north wall of the Access Area. We also indicate the maneuvering clearance, 50″ wide × 60″ deep, on the Access Area side of the door. It is unnecessary to indicate maneuvering clearances on the Lobby side of the door, because we can readily see that a 44-inch-wide × 48-inch-deep floor space exists, and no further development on this side of the door will affect circulation.

The Landings

We now have a door joining Building One with the Access Area, but the spaces on opposite sides of the door are at two different floor levels. Therefore, we must create a landing in the Access Area at the same floor level as Building One. At this early stage, we decide to extend the landing 18 inches from the latch side of the door (coinciding with the maneuvering clearance) and run the landing across the entire north end of the Access Area. We note this floor elevation as 18″, the same as the floor level of Building One.

With a difference of elevation between floors of 18 inches, and a maximum ramp slope of 1:12, we know it will take 18 feet of ramp to affect the vertical transition. However, the distance that remains between the north landing and the south side of the Access Area is less than 20 feet. That dimension will not accommodate 18 feet of ramp, plus a handrail extension of 12 inches, plus the required 5-foot level landing at the bottom of the ramp. Thus, it appears that

our ramp will require at least two separate runs. Using a switchback ramp requires an intermediate landing, and this we place along the south wall of the Access Area. This landing is shown 60 inches deep, which is the minimum turning radius of a wheelchair.

The Ramp

The program requires that all ramp widths be no less than 44 inches, and that is the width we use to create our first layout. If we scale the first leg of the ramp, drawn as described in Figure 12.4, we find that its length is about 13.5 feet. Therefore, the final leg of the ramp would be about 4.5 feet, for a total of 18 feet. We can also see that the short length of ramp ends a couple of feet before the existing doorway. At this point we are certain that a 1:12 ramp slope will work, within the space provided, for the 18-inch floor height differential. Before going any further, however, we must be sure that the stairway will fit as well.

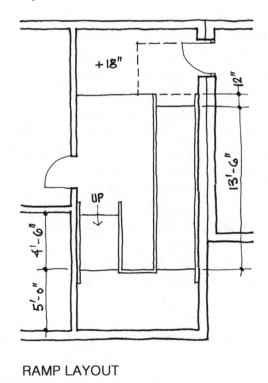

RAMP LAYOUT

Figure 12.4

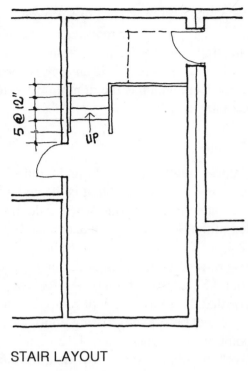

STAIR LAYOUT

Figure 12.5

The Stairway

With the elevation differential of 18 inches, we decide to use six-inch risers for the stairs, which results in exactly three risers. The stairs must be placed between the existing door at the Building Two Lobby and the north landing in the Access Area. Three risers produce two treads (always one less than the number of risers), and these are drawn with a depth of 12 inches and the minimum prescribed width of 44 inches. The exact location of the stairs is determined by measuring 12 inches for maneuvering clearance from the latch side of the Access Area door, another 12 inches for the handrail extension, and then the two 12-inch treads plus another 12-inch handrail extension.

Before completing the details of this vignette, we should determine if this arrangement is the best one possible. Does it satisfy the requirements of function and efficiency? Even though the problem is essentially solved at this point, we can review the arrangement of elements to

see if a more efficient layout is possible. One concern here is that the two ramp widths of 44 inches each do not fully occupy the approximate 10-foot-wide Access Area. Some space will be left over, and the disposition of that space should be considered.

We have illustrated in Figure 12.6 three possible arrangements. The first of these places the two ramp segments adjacent to each other and against the east wall of the Access Area. It is an efficient arrangement, but the leftover space is long, narrow, and relatively unusable. In the second scheme, adjacent ramp segments are placed against the west wall of the Access Area, and the remaining open space is available for planting or display. The problem here, however, is that there is insufficient width at the bottom of the ramp where the handicapped user must turn to enter the Lobby of Building Two. The width of this space is the same as the ramp, 44 inches, whereas a wheelchair requires a turning radius of 60 inches. Thus, we can eliminate Scheme 2. The last scheme is the first one developed, and the leftover space here is placed between the two ramp segments. This scheme resolves all the functional problems, and the leftover space does little more than make the arrangement appear more spacious.

We select the arrangement in the last scheme, because we feel that it has the fewest problems. In general, the differences among alternative designs may be slight, and other solutions may be acceptable. In fact, there is still another alternative that has not been considered. That alternative is to make the ramp and stairs wider in order to fill the entire width of the Access Area. The specified 44-inch widths are minimums, and they could be somewhat wider. Candidates should be cautioned, however, that it is generally best to use minimum allowable widths, because excessive widths might be considered inefficient design. Nevertheless, we show that alternative in Figure 12.7.

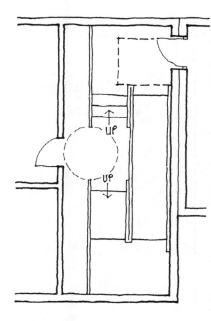
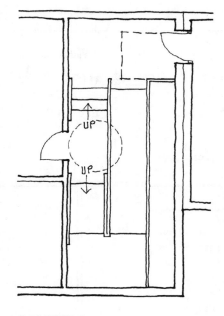
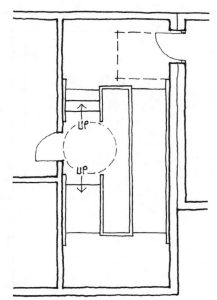

SCHEME 1   SCHEME 2   SCHEME 3

Figure 12.6

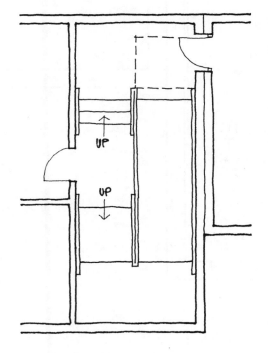

5' WIDE RAMP & STAIR

Figure 12.7

Completing the Design

We must now complete our vignette with the addition of required handrails. All the direction needed is contained in the program. Handrails are required on both sides of ramps and stairs, they must be continuous, and where not continuous, they must extend 12 inches beyond each run. We add the necessary handrails, note the elevation of all landings, and our work is done.

You should be certain that the minimum handicapped turning space can be inscribed at both ends of the ramp, even though it is not required to be shown. A five-foot-diameter circle represents the minimum amount of space necessary to maneuver a wheelchair, and it must be available wherever a wheelchair is required to make a turn.

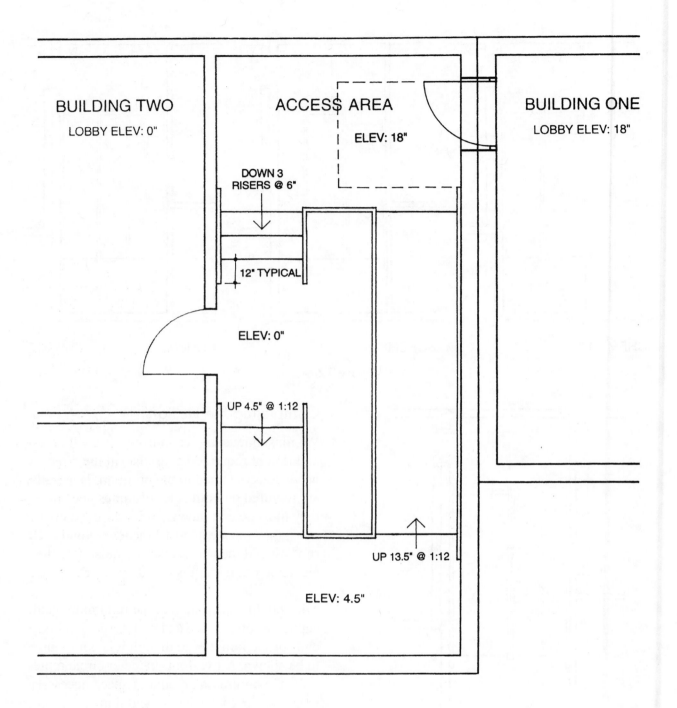

BUILDING TWO
LOBBY ELEV: 0"

ACCESS AREA
ELEV: 18"

BUILDING ONE
LOBBY ELEV: 18"

DOWN 3
RISERS @ 6"

12" TYPICAL

ELEV: 0"

UP 4.5" @ 1:12

UP 13.5" @ 1:12

ELEV: 4.5"

FLOOR PLAN

ACCESSIBILITY—RAMP VIGNETTE SUGGESTED SOLUTION
Figure 12.8

ROOF PLAN VIGNETTE

13

INTRODUCTION

The Roof Plan vignette tests your understanding of the basic concepts of roof design. It also tests your ability to create a roof that effectively drains water from its surface. Candidates are given a background floor plan of a small, relatively simple structure that indicates the outermost roof edges. Candidates must then complete the roof plan by indicating the direction and pitch of roof slopes, noting spot elevations at key points on the roof, and locating roof equipment and accessories to satisfy the requirements of the program. Roof plan solutions are analyzed and scored for completeness and compliance with the program requirements.

VIGNETTE INFORMATION

The Roof Plan vignette is presented on several screens, which are accessed from the index screen and organized as follows:

■ **Vignette Directions**—This screen contains instructions for providing a roof plan that will effectively remove rainwater from all roof surfaces by means of roof slope, gutters, and downspouts. Candidates must define, for each roof area, the extent, slope, and elevations at ridges and eaves. You must also locate on the roof any or all of the following: HVAC condenser, plumbing vent stacks, exhaust vents, skylights, flashing, crickets, and location of a clerestory window.

■ **Program**—This screen describes a small building for which you must prepare a roof plan. The building consists of two volumes, one higher than the other, which have roof heights and slope requirements that differ from each other. Other instructions that will influence your design are organized in the subdivisions listed below.

ROOF DRAINAGE

■ Only roof slope, gutters, and downspouts may be used to remove rainwater.

■ Downspouts may not conflict with any door, window, or clerestory window.

- Rainwater may not run off the edge of any roof or gutter to a lower roof or to the ground. (In other words, downspouts must be used to carry water downward.)

CONSTRUCTION

- Roof sections generally have no overhangs or eaves.

- The minimum ceiling height is given.

- The thickness of the roof and structural assembly is given.

- A range of roof slopes is given for both the high and low building volumes. For example, *the roof slope over the office wing shall be between 2:12 and 5:12.*

- If a clerestory window is included, it is noted here.

- Skylights must be provided at spaces without windows or clerestory, except for corridors, storage rooms, and similar spaces.

- Flashing must be provided where roof and wall surfaces intersect.

MECHANICAL

- The location of the HVAC condenser is based on its proximity to roof edges and clerestory glass, as well as the slope of the roof on which it is set.

- Vents must be provided for exhaust fans and plumbing fixtures.

- **Tips**—This screen contains suggestions that should help candidates to develop a solution more quickly, such as the following examples:

  - To view the roof limits only, use the *layers* tool to turn off the floor plan display.

  - The *check* tool tells you if your roof planes have been drawn within the given limits.

- The *zoom* tool should be used where roof planes meet, since roof edges must coincide.

- The *set roof tool* is used to establish the direction and amount of roof slope, and the required spot elevations.

- The *cursor* and *ortho* tools will help align roof planes with a given roof edge.

- No roof construction may project beyond the dashed roof limit lines, except for gutters and downspouts.

The **work screen**, on which you are required to draw your solution, may be accessed while viewing any of the screens just described. This is done by pressing the space bar, and you may return to any of the information screens with another touch of the space bar. The work screen displays the building floor plan in light gray lines, while the roof limit lines are shown with black, dashed lines. No new roof surface is permitted to extend beyond the dashed lines. If a chimney penetrates the roof surface, its outline is shown with a solid black line. Also shown are exterior dimensions of the building wings, so that you can correctly locate, for example, a ridge line at the center of a gabled roof.

DESIGN APPROACH

A click of the *draw* tool reveals a lengthy list of elements that must be incorporated into the roof plan. Before drawing any of these elements, however, a candidate must have some general idea of how the roof areas will slope in order to shed water. Furthermore, all roof areas should be completed before adding any other element, such as gutters and downspouts, to your design.

It is usually best to begin the roof design at the low portion of the building because the height restrictions in the program are generally explicit. For example, if the minimum ceiling height is given as 8 feet and the thickness of the roof's structural assembly is given as 2 feet, then the lowest roof elevation at the wall line (which coincides with the roof limit line) will be exactly 10 feet. From this starting point, the roof surface will slope upward, within the prescribed range of slopes, to a ridge or roof boundary.

When a roof surface is drawn, roof elevation markers (question marks), slope direction markers (arrows), and slope value markers (question marks) appear. You must then click on the *set roof tool,* which allows you to select the proper values that replace arrows and question marks. You are advised to do this at the time each roof portion is drawn, so that you do not forget what you originally had in mind. In addition, a certain amount of computation is required. For example, if a 32-foot-long roof section begins at elevation 10′-6″ and slopes at a rate of 3:12, it will rise to an elevation of 18′-6″. This condition is illustrated in the next column and calculated in the following way: For every 12 feet (measured horizontally, not along the slope) the roof surface rises 3 feet. In 32 feet, therefore, the roof rises $(3 \div 12) \times 32$ feet = 8 feet. Adding 8 feet to the starting elevation of $10' \times 6''$ results in an elevation of 18′-6″.

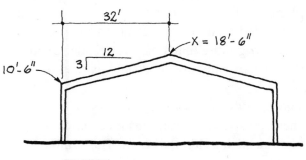

SECTION

DETERMINING THE RIDGE ELEVATION

Figure 13.1

PROBLEM AREAS

The Roof Plan vignette has several problem areas that candidates might avoid with some planning. To begin with, if you have a general roof drainage plan in mind before starting and aim for a simple plan, you will avoid creating complex geometric relationships. In addition, you must be certain that adjacent roof planes align. Every roof plane need not be a single-sloping shed roof, but avoid creating a complex series of hips, valleys, and ridges that take a great deal of time to develop. If the floor plan is symmetrical, that generally suggests gabled roofs with ridges at the center.

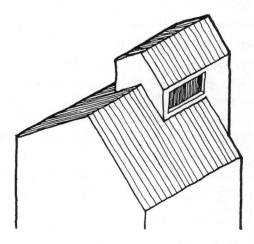

CLERESTORY WINDOW

Figure 13.2

Another problem area is providing sufficient space between high and low roofs to accommodate a clerestory window. You must be certain that the given clerestory height is maintained between the low point of the high roof and the intersecting roof below. This is best accomplished by placing the clerestory window between two parallel roof edges, if the design allows that configuration.

GUTTERS AND DOWNSPOUTS

When all the roof sections are drawn and slopes and elevations are established, add the required gutters and downspouts. Gutters applied to the lower edges of roofs are the only means permitted for removing rainwater, since water is not permitted to drain off a roof edge. Water is also prohibited from draining directly off a gutter onto any lower surface. Downspouts are required to drain gutters to roofs at a lower elevation or to the ground. The principle here is that rainwater must not be allowed to flow in an unrestrained way, which might cause splashing, discoloration, erosion, or other damage to a structure.

Gutters and downspouts are two of the elements found under the *draw* tool, and they are easily applied using a mouse. Gutters less than about 15 feet long require a downspout at one end, while longer gutters should have downspouts at each end. You should be certain that downspouts do not run in front of any door, window, or clerestory, and placing them at the ends of gutters will generally assure compliance.

CLERESTORY AND FLASHING

Clerestory windows and flashing are two more elements found under the *draw* tool. The *clerestory* indication defines the limits of the required window, while *the flashing* indication indicates the intersection of a roof and wall, or roof and chimney. Both indications are similar in that they are applied with a mouse and appear as thick lines; clerestory is a yellow line, while flashing is shown as an orange line. One additional and similar element is a *cricket*, which may be necessary to place at the upslope side of a chimney that penetrates a roof surface. A cricket is drawn with the mouse by

first establishing its base along the chimney surface and then creating the triangular-shaped element.

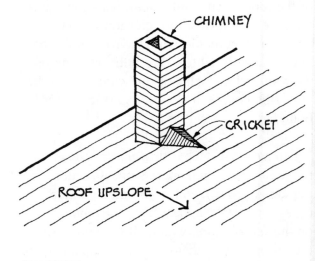

CRICKET

Figure 13.3

MECHANICAL ELEMENTS

The mechanical elements generally incorporated in the Roof Plan vignette are plumbing vent stacks, exhaust fan vents, and roof-mounted HVAC condensing units. Each of these elements is found under the *draw* tool icon on the work screen, and each is relatively simple to apply to the roof plan. Plumbing vent stacks are required at every wall having a plumbing fixture, and the vent stack symbol is centered directly above all such walls. At a wall containing more than one fixture, such as in a toilet room, only one vent stack is required.

The exhaust fan vent symbol is used in any enclosed space that requires mechanical ventilation, such as a toilet room. The symbol may be located anywhere within the space. The location of a roof-mounted condensing unit is generally restricted by its proximity to roof edges, as well as the roof slope on which it sits. For example, the program may specify that such a unit may be placed *no closer to a roof edge than 4 feet and on a roof with a slope of 4:12 or less.* Candidates should place the condensing unit in a central location and preferably over a space that will not be affected by noise or vibration, such as a closet or hallway. In addition, a condensing unit should never be placed in front of a clerestory window. Concern for shielding the condenser from view from ground level is not typically a part of this exam.

SKYLIGHTS

The requirement for natural light in all rooms is generally satisfied by means of windows, a clerestory window, or skylights. If no window is shown in a space, and a clerestory window has not been specified, provide a skylight (unless the space is strictly utilitarian, such as a closet or hallway). As with other elements found under the *draw* tool, a click of the mouse places the predetermined skylight shape at the location selected.

KEY COMPUTER TOOLS

- **Layer Tool**—With the multiple layers present in this drawing, the layer tool will become a helpful tool to distinguish between them.
- **Zoom Tool**—This tool will help the candidate be certain that the roof plane is placed within the building envelop.
- **Full Screen Cursor**—This tool will also help align elements between layers.

VIGNETTE 2 ROOF PLAN

Prepare a roof plan for the small commercial building shown in the plan. The main access street is located to the east, and a service road is located at the west. A high roof covers the Reception and Conference areas, while a lower roof covers all other areas. There are no overhangs; the outermost edges of roof planes coincide with exterior walls, except where indicated at the two exterior doors. Establish and indicate the maximum building height and show roof elevations at the eave and ridge of each roof plane.

Program Requirements
Roof Construction

- All roof areas have a positive slope, and the maximum building height is 20 feet.
- The roof over the Reception and Conference area shall slope no less than 4:12.
- All other roofs shall slope no less than 2:12. Indicate all roof slopes.
- The finish floor elevation is 0'-0", and the minimum ceiling height is 8'-6".
- The roof and structural assembly thickness at all areas is 18".
- The Conference Room has a continuous clerestory window 24" in total height facing south. Indicate extent of clerestory, as shown in the legend.
- Provide 24" square skylights to light the toilet rooms and corridors.

Roof Drainage

- Remove rainwater from all roof areas with gutters and downspouts.
- Downspouts must not conflict with doors, windows, or clerestory window.
- Rainwater must not run from the edge of a roof or gutter onto a lower roof or ground without a downspout.
- A downspout is required for every 450 square feet of roof surface.

Mechanical

- Locate the 3-foot-square HVAC condensing unit a minimum of 4 feet distant from all roof ridges and eaves of the roof surface on which it is located.
- Provide one exhaust fan vent for each toilet room.
- Provide plumbing vent stacks where required to vent plumbing fixtures.

LEGEND

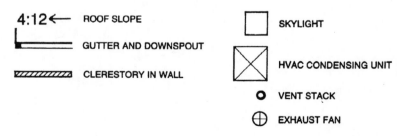

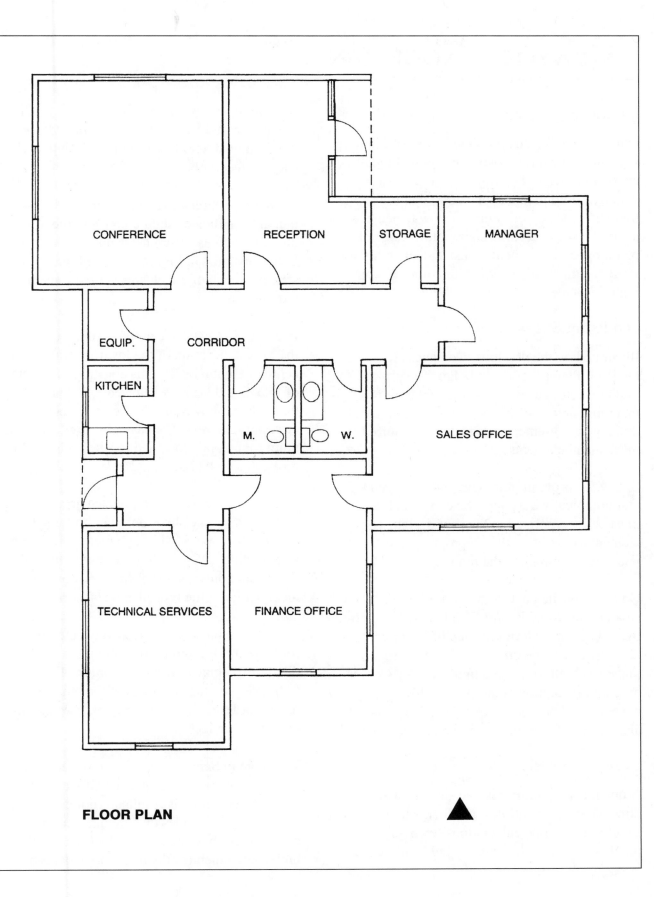

CONFERENCE

RECEPTION

STORAGE

MANAGER

EQUIP.

CORRIDOR

KITCHEN

M.

W.

SALES OFFICE

TECHNICAL SERVICES

FINANCE OFFICE

FLOOR PLAN

VIGNETTE 2 ROOF PLAN

Introduction

Our example for this problem is a Roof Plan vignette created for a paper-and-pencil mock exam. Following the program, we present a detailed solution and explanation that illustrate the process you should follow to achieve a passing score. Roof design requires logical planning, careful calculations, and an ability to visualize three-dimensional junctions of sloping roof planes.

The Exam Sheet

Illustrated on the previous two pages are the Roof Plan program and plan. The program describes a small commercial building, including details of its roof construction, roof drainage requirements, and a list of required mechanical elements.

Below the program requirements is a legend that graphically describes the elements that must be incorporated in the roof design. These elements, together with the program, are all you need to complete the required roof plan.

On the opposite page is the floor plan for a small commercial building. Since there are no overhangs, the exterior outline of the building also represents the outermost edges of roof planes. A high roof is required over the Conference and Reception areas, while a lower roof covers all other areas of the plan. The two roofs also have different minimum slopes.

Design Analysis

A quick review indicates that several plan dimensions are multiples of 6 feet, which should make slope calculations relatively simple. For example, the high-roofed Conference/

Reception area is 18 feet wide, and so, too, is the low-roofed Sales Office area. In addition, the Technical Services and Finance Office are each 12 feet wide.

There are several ways to roof this building, and candidates should consider the more obvious choices before determining a solution. Illustrated in Figure 13.4 are four distinct schemes, each of which solves the problem requirements.

In every case, we have assumed that 10 feet is the lowest elevation of the low roof (minimum ceiling height of 8'-6" plus construction depth of 1'-6"), and we have allowed 3 feet between the high and low roofs for the two-foot-high clerestory window. We have also used the minimum roof slopes specified for the high and low roofs, that is 4:12 and 2:12, respectively.

Scheme 1 is the least complicated arrangement and uses a simple shed roof over both the high and low roof areas. Although there is little logic in providing a 17-foot-high roof over the Sales Office, and only a 10-foot-high roof over Technical Services, the roof plan does work.

Scheme 2 retains a simple shed for the high roof, but a ridge is provided along the wall dividing Technical Services from the Finance Office. Again, the two sides of the building have roof levels that are different, but this time, by only three feet.

The ridge in Scheme 3 has been moved over the wall separating the Finance Office from the Sales Office, and this results in roof eave elevations at opposite sides that are different by only one foot. We have also created a ridge for the high roof, which results in a more balanced

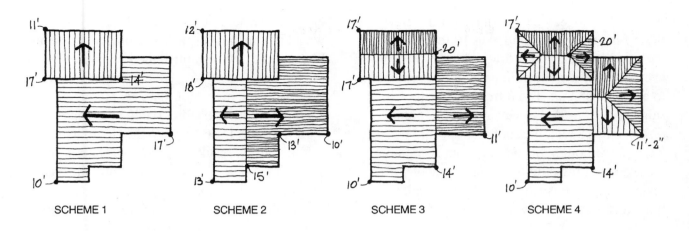

SCHEME 1   SCHEME 2   SCHEME 3   SCHEME 4

ALTERNATIVE ROOFING SCHEMES

Figure 13.4

form. The total building height in this scheme reaches the limit of 20 feet at the highest ridge.

Scheme 4 is presented to show candidates how unnecessarily complex a roof plan can become. The high roof was designed as a relatively simple hip roof, but the low roof, which attempts to be consistent, has become a terrible mess. There is another hip roof over the Manager and Sales Office, but the remainder of the low roof design simply falls apart. The shed roof at the west side is inconsistent with the hip roof at the east. The hip roof would butt into a vertical wall, and the two sections would appear as two distinct buildings. Candidates are advised to avoid complex solutions such as this. There is never enough time to resolve all the conflicts; the slope calculations take much more time because there are so many individual roof sections, and there is little to gain from the additional complication.

In conclusion, Scheme 1 is simplistic, Schemes 2 and 3 are acceptable, and Scheme 4 should be avoided at all costs.

Developing the Low Roof

Once you have a general idea of how water will be drained from the roof, you must verify roof slopes and calculate key elevations. In a problem such as this, where a maximum building height is given, begin with the maximum height at the highest point of the high roof (20 feet), or one may begin with the lowest elevation of the low roof. You may recall that the lowest elevation, 10 feet, is the sum of the minimum ceiling height (8'-6") and the thickness of the structural roof assembly (1'-6").

We begin by following Scheme 3 and establish the low point of the low roof, 10 feet, along the west wall. From that side, we rise at the slope rate of 2:12 (that is, 2 feet vertically for every 12 feet horizontally) in an easterly direction to the wall dividing the Finance Office from the Sales Office. In that horizontal distance of 24 feet, the roof will rise 4 feet to 14 feet at the ridge. From the 14-foot-high ridge, we drop to the east wall at the same slope rate of 2:12. The east wall is 18 feet distant, and thus, the roof elevation at the easternmost eaves is 11 feet (14'-3').

As an alternative, we could have set the elevation of the eastern roof edge at 10 feet to match the opposite side of this low, gabled roof. However, the roof slope would have to be modified to accomplish that. To go from the 14-foot-high ridge to a 10-foot-high eave, in a horizontal distance of 18 feet, the roof slope would be 2.67:12. Although this would be an acceptable solution, and it would have the advantage of maintaining a consistent eave line, the additional calculations may not be worth the time spent.

Developing the High Roof

The lowest elevation of the high roof is established at a point that allows space for the clerestory window in the south wall of the Conference Room. The clerestory requires 2 feet of total height, which includes the window frame, glass, flashing, etc. If we measure westward from the ridge of the low roof to the beginning of the Conference Room, we can determine that the roof elevation at that point is 12 feet (14'-2'). This is the critical elevation, because the roof continues to slope downward and provides even more vertical space for the clerestory.

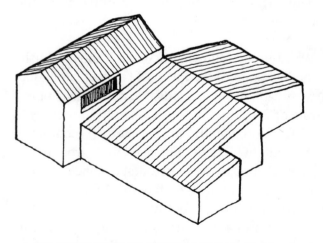

LOCATING THE CLERESTORY WINDOW

Figure 13.5

If we provide 2 feet of clerestory space above the roof elevation of 12 feet, the top of the

clerestory would be at 14 feet. From this level, we add 18 inches for the structural roof assembly, and thus, we establish the low point of the high roof as 15'-6". We have established a ridge over the high roof that is centered on its 18-foot width. Therefore, rising at a slope of 4:12 from the 15'-6" eave to the ridge, which is 9 feet away, results in a ridge elevation of 18"-6" (15'-6" + 3'-0"). At this point, the basic roof design is complete, and the slopes and elevations are set. We next move on to the other requirements of this vignette.

Adding Gutters and Downspouts

Rainwater is not permitted to run off the edge of a roof or a gutter to any lower surface without using a downspout. First, however, we must apply gutters to the low edge of every roof section. These are applied to the east and west sides of the low roof and the north and south sides of the high roof. Two downspouts are added along each gutter, since the shortest gutter is about 28 feet long, and more than one downspout is generally necessary when the gutter length exceeds 15 feet. Downspouts are typically located at the corners of the building.

There is another restriction in the program that must be considered: One downspout is required for every 450 square feet of roof surface. The high roof poses no problem, because each half has two downspouts that drain only about 250 square feet. The eastern section of the low roof also has two downspouts that drain about 500 square feet of roof. The western section of the low roof, however, drains 840 square feet of roof. In addition, half of the southern portion of the high roof also drains onto it by means of the downspout in the southeast corner of the high roof. This roof adds another 125 square feet or so, making the total roof area drained 840 + 125 = 965 square feet. Therefore, 965 square feet ÷ 450 square feet per downspout = 2.14, and therefore, three downspouts are required

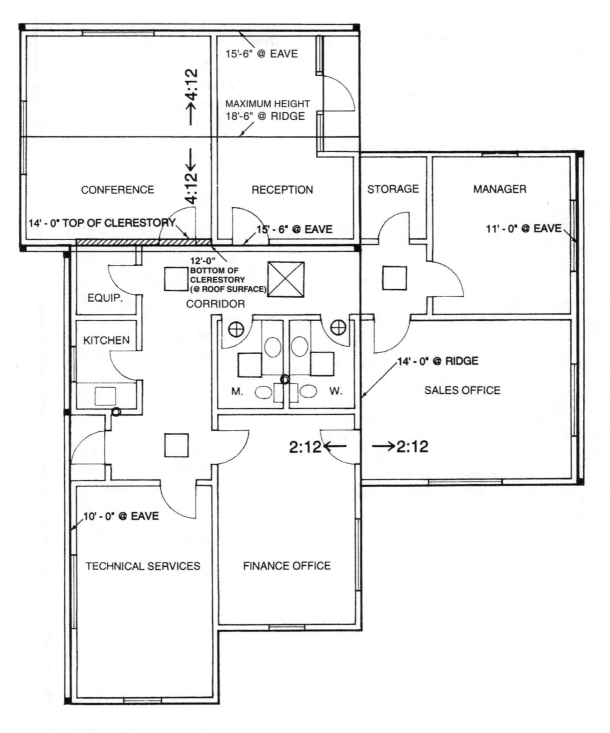

FLOOR PLAN

ROOF PLAN VIGNETTE SUGGESTED SOLUTION

Figure 13.6

for this section of roof. We locate the third downspout just north of the doorway opening to the Corridor.

Adding Skylights

Skylights are required to light the Toilet Rooms and Corridors, and a two-foot square symbol is shown in the legend. Since we are not told how many skylights to use, use common sense. One skylight is centered in each of the Toilet Rooms. For the Corridors, we have placed one skylight in front of the doorways at each end and another skylight midway between the two, where the Corridor changes direction. These should provide ample light for circulation.

Adding Mechanical Elements

The largest of the remaining elements to locate is the 3-foot-square HVAC condensing unit. The only restriction here is that it must be 4 feet distant from all ridges and eaves of the roof surface on which it is placed. We locate the unit over the Corridor midway between two skylights. It is a fairly central location, and at that site, it should minimize any disturbance from noise or vibration.

We next locate the required exhaust fans in the Toilet Rooms. These may be shown anywhere within the spaces. The vent stacks, on the other hand, must be centered over the walls against which plumbing fixtures are placed. We place one vent stack symbol in the plumbing wall shared by the two Toilets and another at the sink in the Kitchen.

Conclusion

Because Roof Plan vignettes must be completed in such a short time, candidates can do little more than develop their first workable idea for a roof plan. There is not enough time to try a variety of alternate schemes or worry about details. Much of the work of this vignette consists of determining the direction and amount of roof slopes and calculating roof elevations. Work carefully, because the program will catch every numerical mistake. Perform all calculations with the on-screen calculator.

Work systematically to solve the large problem first and then add the necessary details later. The omission of a vent stack will not carry the same penalty as the failure of adjacent roof planes to align. Keep an eye on the big picture and avoid suffering over minor details.

STAIR DESIGN VIGNETTE

INTRODUCTION

The Stair Design vignette tests your understanding of the three-dimensional nature of stair design and of the basic functional and code issues involved. The candidate is provided with upper and lower floor plans, a program, code requirements, and a reference building section. You are then required to complete the plans with a stair system that satisfies the programmatic and code requirements. Among the typical elements a candidate must provide are stairs, landings, handrails, guardrails, an area of refuge, and landing elevations. Solutions are analyzed for completeness and compliance with program and code requirements.

Candidates have historically found this to be one of the more difficult vignettes. This is likely due to the number of components to simultaneously track: treads and risers, allowable head room, setting the elevations, and often providing an area of refuge.

VIGNETTE INFORMATION

The Stair Design vignette begins with the Task Information Index screen. This screen allows access to a series of other screens that contain information necessary for solving the stairway problem. The available screens are listed below.

- **Vignette Directions**—These instruct candidates to provide an exit stairway in an existing two-story stairwell that is shown on the work screen. You are directed to comply with all code requirements and, among other provisions, to indicate the elevations of all landings.

- **Program**—This repeats the basic instruction to design a stairway as a means of egress from separate floor levels leading into the existing stairwell and ultimately to an exterior exit at grade. The exit from the uppermost level must be a continuous path to the ground floor exit with a landing at an intermediate level. Candidates are informed about the stair construction and the landing and stringer depths, which allows them to determine headroom.

- **Code**—These excerpts are the only code-related criteria a candidate is permitted to use. The requirements are organized under the following headings:

  - Definitions—including *means of egress, exit stairway*, etc.

  - Capacity of Exit Components—including occupant loads and minimum allowable width.

  - Stairways—including width, headroom, landing width, and riser and tread dimensions.

  - Doors—including reduced landing width and required floors on each side.

  - Guardrails—at open sides of landings.

  - Handrails—including continuity, projections into passageway, and extensions beyond top and bottom of stair runs.

  - Area of Refuge—including size and location.

- **Section**—The drawing on this screen is cut through the stairwell and indicates the elevations of all floor levels leading into the stairwell.

- **Tips**—This screen suggests a number of strategies to make work more efficient. Some examples follow:

  - The stairway should be designed in the same order as the tasks are presented.

  - The number of risers required should be calculated before stairs are laid out.

  - As a stair flight is drawn, tread depth is indicated at the bottom of the work screen.

  - A question mark represents a request for an exact elevation, which must be calculated and noted on the drawing using the elevation tool. The elevation must be indicated for both the stairs and the land, even if the elevations are the same.

  - Areas without development are assumed to be open to the spaces below.

  - Different floor levels are accessed by using the *layers* tool.

  - Fine adjustments require use of the *zoom* tool.

While viewing any of the screens described above, you can tap the space bar to access the **work screen**. This screen shows both floor levels of the stairwell. One level is shown with black lines and the other with gray lines. One may draw on the plan shown in black, but not on the plan in gray. However, you can toggle between the two plans with the *layers* tool, which changes the gray floor plan to black, showing you to draw on it.

DESIGN APPROACH

Designing a stairway is a fairly mechanical process. Because of human size and restrictions of physical movement, the range of riser heights is relatively small, and it is limited in this vignette to between four and seven inches. The relationship of riser height to tread depth is critical and permits little variation. It is proportioned to accommodate safe and comfortable body movement and is always based on having risers and treads of uniform size. Without uniformity and proper proportions, a stairway might be difficult and dangerous to use. Several rules-of-thumb have evolved over the years to express a proper riser to tread ratio, such as, 2 risers + 1 tread = 24 to 25 inches. Thus, with a riser height of six inches, the tread depth would be between 12 and 13 inches [24 or 25 − (2 × 6) = 12 or 13].

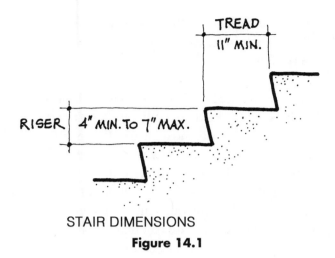

STAIR DIMENSIONS

Figure 14.1

Other details to keep in mind are as follows:

- In any stair flight, the number of risers is always one more than the number of treads. For example, a flight consisting of 10 risers would contain 9 treads.

- The number of risers is determined by dividing the total vertical rise in a flight by the desired riser height and rounding up the result to a whole number. For example, if the floor-to-floor height is 10'-4", and the desired riser height is 7", the number of risers required would be $[(10' \times 12") + 4"] \div 7"$ risers = 17.71, which is rounded up to a required 18 risers.

- The least landing dimension is never less than the required stairway width.

- The minimum headroom of all parts of a stairway is not less than 80 inches.

- Handrails are continuous on both sides of a stairway, and at least one handrail must extend horizontally 12 inches beyond the top and bottom risers.

DESIGN PROCEDURE

Clicking on the *draw* tool at the upper left part of the screen brings up the elements necessary to include in your vignette solution. These are

listed as **Railing, Landing, Stairs** (one partial flight shown in the north-south direction and another in the east-west direction), and **Cut Stairs**, which are examples of stair runs (shown in every possible direction) with cut lines drawn halfway through them. One should begin a stair layout by establishing either the highest landing in the stairwell or the lowest. Before establishing a landing, however, one must determine the landing's dimensions.

THE LANDING

Since the least dimension of a landing may not be less than the required width of the stairway, you must first determine the stairway width. The program often lists the occupant load for each building level together with the number of exits from each of these levels. For example, the program may state that the occupant load for the second floor is 320 and the number of exits from that floor is 2. The occupant load for each exit is determined by dividing the total occupant load for the entire floor by the number of exits serving that floor. In this example, therefore, each exit will accommodate 160 people (320 ÷ 2). Next, the width of each exit component in inches cannot be less than the occupant load of each exit multiplied by 0.3, which is a constant found in the Stair Design vignette code. The minimum width, however, may not be less than 44 inches. In the example, the minimum width would be 160 × 0.3 = 48 inches. Candidates need not memorize any of the above factors, because they will be given in the vignette code requirements.

This upper landing should be at the same elevation as the second floor level. In other words, there must be no change in elevation at the door threshold that separates the landing from the existing floor level, because this would create a hazardous condition. With the minimum dimensions and elevation established, one may proceed to lay out the uppermost landing.

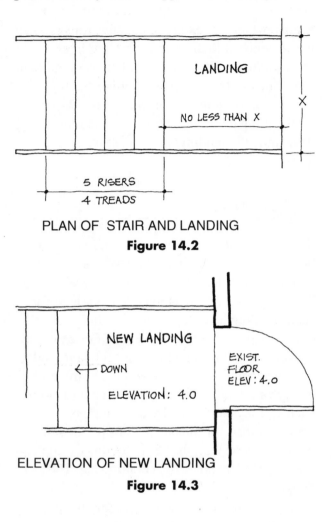

PLAN OF STAIR AND LANDING
Figure 14.2

ELEVATION OF NEW LANDING
Figure 14.3

THE STAIRWAY

Before establishing the first flight of stairs that descend from the uppermost landing, a candidate must determine where the next lower landing will be located. To do this, you should have a rough notion of how the stairway will eventually reach the target elevation, which is the ground floor exit to the exterior. Somewhere between these two points will usually be an intermediate landing at the same elevation as the intermediate opening to the stairwell. In past problems, the floor elevations of the three levels were given as even feet or half feet, sug-

gesting to candidates that a six-inch-high riser would result in an exact number of risers at every stair run.

Once the vertical distance between two floor levels is determined, a candidate must divide this distance by the chosen riser dimension to obtain the required number of risers needed between floor levels. Next, determine the tread width appropriate for the riser height selected. Any of the following conventional standards may be used, and all are calculated with 6-inch-high risers and 12-inch-deep treads:

- 2 risers + 1 tread = 24 or 25 inches
  ($2 \times 6 + 12 = 24$)
- Riser + tread = 17 or 18 inches
  ($6 + 12 = 18$)
- Riser × tread = 70 to 75 inches
  ($6 \times 12 = 72$)

Using the NCARB program, a section of stairs is drawn in the following way:

1. Under the *draw* tool, select *stairs.*
2. Select a stair section that runs either north-south or east-west.
3. Select the number of risers needed between two landings.
4. When placing the stairs, a note on the screen indicates the variable tread depth as the stair section is lengthened or shortened. Simply stop when the desired number is reached.

SETTING ELEVATIONS

When you establish a landing or a stair section, you will see an on-screen question mark in the middle of the landing or at the top of the highest tread and bottom of the lowest riser. These question marks are a request for you to indicate the elevations at those levels. You first click the *set elevation* tool, then click on the question mark, and this brings up a box in which the elevation is noted. One mouse click later, the question mark is replaced by the elevation.

We suggest that you establish and set elevations as you go along, because as you draw elements, you are aware of their elevations. If you wait until the several stair sections are laid out, you may forget the earlier calculations you performed.

Jotting the calculations on the scratch paper provided will also be helpful.

Do not be surprised if several elevations are similar, because the top of the highest tread is also the landing from which it starts. Similarly, the elevation of the bottom of the lowest riser is the same as the elevation of the landing from which the riser begins.

HEADROOM

Stair Design vignettes invariably involve problems of headroom. In other words, a problem is rarely solved without one section of stair passing beneath another or under a landing. In those cases, you should be certain to provide sufficient clearance so that a person using the stairs will not bump one's head.

The minimum headroom of all parts of a stairway must not be less than 80 inches measured vertically from a tread nosing, floor surface, or landing. Within the program description of the stair construction, you will find the critical dimensions necessary to determine headroom. For example, the program may state: *Stair sections are 12 inches deep between the stair nosing and the stringer soffit measured perpendicular to the soffit.* With this information, one would know that 12 inches (measured

perpendicular to the soffit) must be added to the 80-inch required clearance (measured perpendicular to the tread) to determine the overall dimension from one tread to another directly above. The depths of landings are described similarly, and headroom clearance between a landing and the soffit of another landing directly above must also be no less than 80 inches.

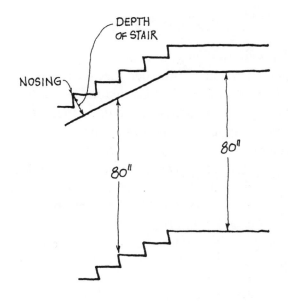

MINIMUM HEADROOM

Figure 14.4

When a candidate uses the *layers* tool to toggle between floor levels, the level selected is shown with black lines and the other level is shown with lighter, gray lines. This shows both floor levels together, making the verification of headroom somewhat easier.

CUT STAIRS

Cut stairs elements, found under the *draw* tool, are used in place of continuous stair elements on the ground floor plan only. The cut stair indication signifies that you would be unable to see the entire run of the stair section, because it has been cut horizontally a few feet above the floor. The blank portion above the cut line represents stairs that continue upward and are out of sight in this plan view. On the other hand, second-floor stair sections are seen in their entirety, because looking down from the second floor, you can see everything that occurs below.

RAILINGS

Handrails and guardrails are added to a solution when you are satisfied that your stair design solves the basic circulation problem. Railings are found under the *draw* tool, and they are applied with a simple mouse click. Rules governing railings are found on the code screen. Essentially, stairways have continuous handrails on both sides, with at least one handrail extending horizontally 12 inches beyond the top and bottom risers. Open sides of landings must also be protected with continuous guardrails.

AREA OF REFUGE

An area of refuge is an open space large enough to accommodate a wheelchair and served by an accessible means of egress. Its purpose is to provide a relatively safe waiting area until fire department personnel are able to physically remove the handicapped person. Areas of refuge are separated from the remainder of the story by a smoke barrier having at least a one-hour fire-resistance rating.

In the Stair Design vignette, a 30-inch by 48-inch refuge area is normally required within the stair enclosure, and it must be located in such a way that required stair and landing widths are not reduced. A 30-inch by 48-inch rectangle may be drawn using the *sketch* tool, but if you allow ample room at the proper location, it will generally be assumed that the requirement is met.

KEY COMPUTER TOOLS

- **Set Elevation Tool**—This tool is used to label elevations at stairs and landings.
- **Layer Tool**—This tool is essential for navigating between floors.
- **Zoom Tool**—This vital tool is especially helpful in this vignette for accurately placing landings and stairs.
- **Calculator Tool**—This on-screen tool will be invaluable for determining stair width based on occupancy.

VIGNETTE 3 STAIR DESIGN

Introduction

This Stair Design vignette appeared on a paper-and-pencil mock exam. It is used here to instruct candidates in an efficient method of solving this type of problem. As stated earlier, stair design is a fairly mechanical process. You are restricted by code and usual design conventions based on human size and needs, and therefore, most stair layouts have relatively few design alternatives. Getting up or down from one level to another must be arranged in as safe, direct, and simple a way as possible. By following the logic of our discussion, you should learn to solve this vignette type with little difficulty.

The Exam Sheet

The program describes the required elements and contains general design demands and specific code standards. Candidates are required to use this information to develop an exit stairway that conforms to all program and code requirements.

Shown are the partial upper and lower floor plans of an open stairwell. Three openings are shown into the stairwell, one from an upper floor Office, another from a Meeting Room at the lower floor, and finally, an exit door at an intermediate level. Candidates should note immediately that the Meeting Room on the lower floor plan is three feet lower than the exterior grade level. There is no indication of a change in floor level, so it must be assumed that the candidate will provide a break in the floor level where appropriate.

Because there is no accompanying section, candidates should consider sketching the existing conditions to better understand the relationships of the floor levels in this problem.

Design Analysis

Several details in this vignette are different and somewhat simpler from the usual computer Stair Design vignette. First, the occupant load for each level need not be calculated, because the stair widths are given. Next, the two stair sections have different widths: one is 44 inches and the other is 60 inches. Finally, the tread and riser dimensions need not be calculated, because they are given as 12 inches and 6 inches, respectively.

The solution to any Stair Design vignette must begin at either the high or low end of the run. In this case, we may begin at the Office door at the upper floor or Meeting Room at the lower floor. Stair runs from both of these spaces must eventually end up near the exterior door at grade. Since the Meeting Room is only three feet below the exterior door at grade, its stair run is relatively short, and this is where we begin.

The Meeting Room Stair

We are told that because of the Meeting Room occupant load, a 60-inch-wide exit path is required. This means that the landing in front of the Meeting Room door must be no less than 60 inches wide as well, because the least dimension of a landing may not be less than the stairway width. Therefore, we measure 60 inches from the outswinging door to represent the clear landing area. The next question concerns the direction of the stairway: should it run north-south, or east-west? Occupants exiting the Meeting Room should be able to walk straight

VIGNETTE 3   STAIR DESIGN

Develop an exit stairway within the existing two-story space shown. Draw all necessary and/or required components including steps, landings, railings, handrails, and indicate elevations of all landings. The completed plan should conform to all program and code requirements.

Program Requirements
General

- A stairway is required as a means of egress from the Office on the upper level and the Meeting Room on the lower level.

- All occupants will exit to the exterior through the indicated exit door at grade.

- The occupant load is such that a 44-inch-wide exit path is required from the Office and a 60-inch-wide exit path is required from the Meeting Room.

Stairways

- The stairway widths shall be as noted above.

- The least dimension of landings shall not be less than the stairway's required width.

- The minimum headroom of all stairway parts shall be not less than 80 inches measured vertically from any horizontal surface.

- All treads and risers shall have a uniform dimension of 12 inches and 6 inches, respectively.

Doors

- Open doors shall not reduce the width of landings to less than ½ the required width.

- A floor or landing on each side of each door shall be at the same elevation.

Rails

- Stairways shall have continuous handrails on both sides.

- Where handrails are not continuous, at least one handrail shall extend horizontally at least 12 inches beyond the top riser and bottom riser.

- Handrails shall not project more than 4 inches into the required passage width.

- Open sides of landings shall be protected by a continuous guardrail.

Area of Refuge

- Provide an accessible area of refuge no less than 30 × 48 inches serving the upper level of the two-story space.

- The refuge space shall not reduce the required passage width.

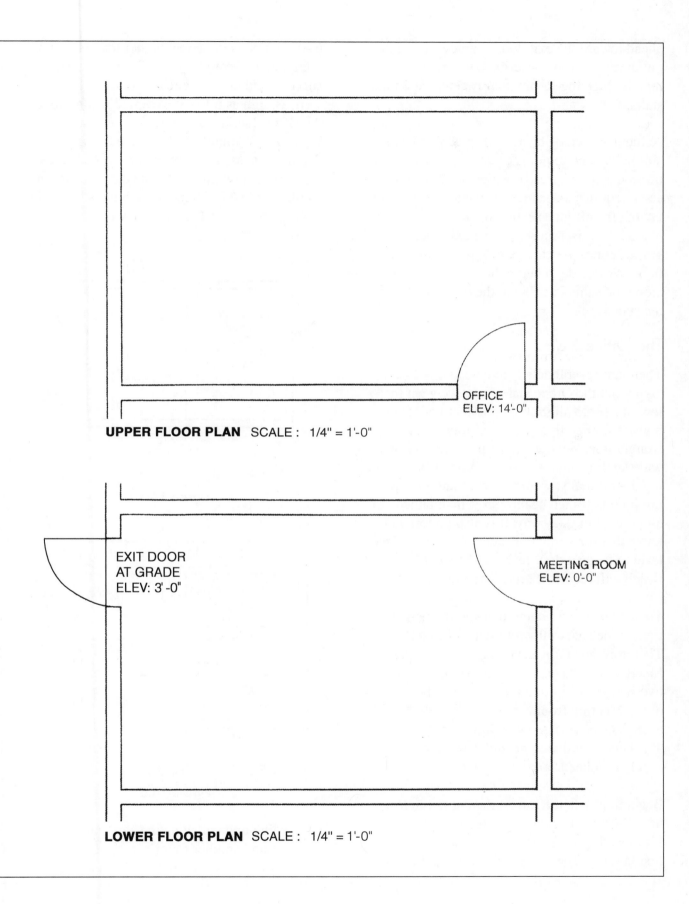

UPPER FLOOR PLAN SCALE : 1/4" = 1'-0"

OFFICE
ELEV: 14'-0"

EXIT DOOR
AT GRADE
ELEV: 3'-0"

MEETING ROOM
ELEV: 0'-0"

LOWER FLOOR PLAN SCALE : 1/4" = 1'-0"

ahead toward the exit door at grade. If you lay out the stairway as described, you will find that the Meeting Room door is centered on the stairs.

Achieving a grade change of three feet, using six-inch risers, requires exactly six risers. This translates to five treads, which at 12 inches each, requires a distance of five feet. Thus, we lay out the 60-inch section of stairs in a space 5' × 5'. The riser closest to the exit door at grade defines the change of level for the lower floor. We decide to begin the Office stair run along the same line, but at the opposite side of the stairwell.

The Office Stair

There are several details to consider before laying out the Office stair. First, we want the lowest riser to align with the Meeting Room's highest riser so that the lower floor break is a straight line. Next, we must provide a 44-inch-wide landing adjacent to the Office's outswinging door. Finally, we must be certain that the minimum 80-inch headroom is maintained at all parts of the stairway. It is also important to know that the vertical distance from the Office level to the exit door level is 11 feet (14'-3'), which will require 22 risers at 6 inches each.

There is one more detail that should be mentioned: the stairwell width scales 12 feet. Therefore, the Office stairway must overlap the Meeting Room stairway at some point, because two runs of the Office stairway plus the width of the Meeting Room stairway exceeds the available space (2 × 44" = 7'-4" + 5'-0" = 12'-4"). This should alert candidates to a potential problem of headroom.

Beginning with the lowest riser at floor elevation 3'-0", we proceed upward, along the south wall of the stairwell, six risers to the first landing. The elevation of this landing is

6'-0", which is three feet higher than our starting point at elevation 3'-0". There is actually space to add one more riser, but if the landing were as high as 6'-6", the overall headroom to the Office landing above would be reduced to 7'-6" (14'-0" minus 6'-6") less the construction depth. If we assume a minimum of 12 inches for construction thickness, the clear headroom would be 6'-6" or 78 inches, which is less than the required 80 inches.

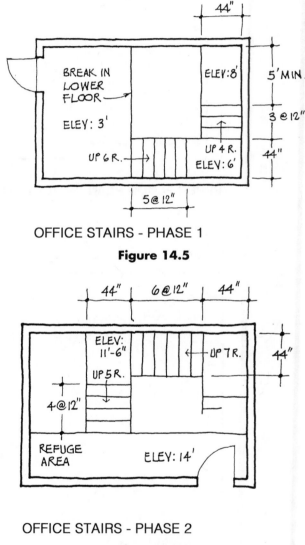

OFFICE STAIRS - PHASE 1

Figure 14.5

OFFICE STAIRS - PHASE 2

Figure 14.6

Proceeding upward along the east stairwell wall, we add another four risers to reach the next landing at elevation 8'-0". This landing is over five feet deep, which avoids encroaching on the landing area in front of the Meeting Room below. It also provides an overall headroom below of 8'-0", less 12 inches for construction, or 84 inches of net headroom.

The remaining six feet of rise begins with seven risers along the north wall and continues to the landing at elevation 11'-6". We then turn southward and complete the run with five more risers to the Office elevation of 14'-0". Along the south wall of the stairwell, we create a continuous 44-inch-wide landing that (1) provides access to the Office, and (2) allows for a 30" × 48" refuge area that is away from the main circulation path. The continuous landing also ensures that any required clearance at the latch side of the door will be satisfied.

Final Presentation

Handrails and guardrails are added only when the candidate is certain that the stairway design has solved all the problems. In our design, the inside handrail is continuous, while the outside handrail is not continuous but extends 12 inches beyond the top and bottom risers of each stair section. Along the extended landing at elevation 14', we provide guardrails that become handrails as they continue down the stair flight towards elevation 11'-6".

At this point we have solved the problem and the stair design is complete. However, to comply with the program requirements, we indicate the elevations of all landings. This is also a good way to verify that the stairway does indeed work. We also add dashed lines on the lower floor plan indicating the outline of the floor above. Finally, we add a cut line on the rising stair section to the Office. This indicates that the stair run continues but cannot be seen in the lower floor plan. Candidates often use stair cut lines on the upper floor plan view, but this is clearly incorrect. When viewed from above, you see everything that occurs below. No part of a stairway is hidden, unless it is blocked from view by another section of stairs.

Conclusion

Stair Design vignettes are relatively conventional problems, but they can be complicated. In this and other similar problems, several alternative solutions might be acceptable, and candidates may have to sketch other possible layouts until an optimum solution appears. However, you have only 60 minutes for this vignette and, therefore, should not waste time seeking the perfect solution.

Candidates might also consider the following useful strategy: Put yourself in the position of an occupant of this facility. Trace the circulation path and mentally feel the rhythm of climbing or descending the stairs. Open the door, hold the handrail, head for the exit, and visualize the experience. You may discover what is both good and lacking in your design.

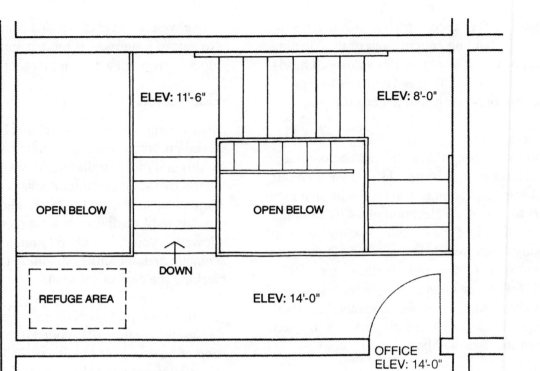

ELEV: 11'-6"   ELEV: 8'-0"

OPEN BELOW   OPEN BELOW

DOWN

REFUGE AREA   ELEV: 14'-0"

OFFICE
ELEV: 14'-0"

UPPER FLOOR PLAN SCALE : 1/4" = 1'-0"

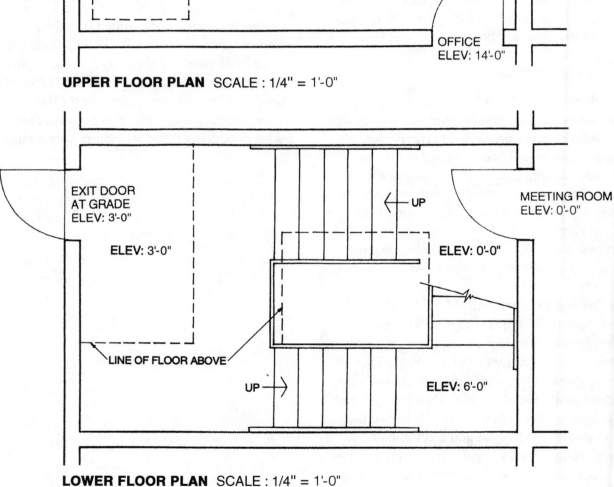

EXIT DOOR
AT GRADE
ELEV: 3'-0"

← UP

MEETING ROOM
ELEV: 0'-0"

ELEV: 3'-0"   ELEV: 0'-0"

LINE OF FLOOR ABOVE

UP →   ELEV: 6'-0"

LOWER FLOOR PLAN SCALE : 1/4" = 1'-0"

STAIR DESIGN VIGNETTE SUGGESTED SOLUTION

Figure 14.7

GLOSSARY

The following glossary defines a number of terms, many of which have appeared on past exams. While this list is by no means complete, it comprises much of the terminology with which candidates should be familiar. You are therefore encouraged to review these definitions as part of your preparation for the exam.

A

Accelerator A substance, such as calcium chloride, added to a concrete mix to speed up its setting and strength development.

Acoustics The science of sound and sound control.

Acrylic A noncrystalline thermoplastic with good weather resistance, shatter resistance, and optical clarity; sometimes used for glazing.

Admixture A prepared substance added to concrete to alter or achieve certain characteristics.

Aggregate The chemically inert element of concrete, usually consisting of sand, crushed rock, and/or gravel.

Air Entrainment The incorporation of tiny air bubbles into concrete or mortar to improve its workability and resistance to freezing.

Alkyd A synthetic resin used as a vehicle for paint.

Alloy A substance generally composed of two or more metals that have been intimately mixed.

Asbestos Cement A material consisting of a mixture of portland cement and asbestos fiber. Although resistant to fire, it is considered a health hazard.

Ashlar A building stone that has been shaped and smoothed into a rectangular form for use in masonry construction.

Asphaltic Concrete A mixture of asphaltic cement and aggregates, used as a paving material, which is spread and rolled while hot.

ASTM American Society for Testing and Materials.

B

Backfill Earth that is replaced around a foundation or retaining wall after the concrete forms have been removed.

Balloon Framing A method of framing wood stud walls in which the studs are continuous for the full height of the building, which is usually two stories, with the joists bearing on a ribbon let into the studs.

Baluster A vertical support for a handrail.

Balustrade A complete railing system, consisting of a top rail, supporting balusters, and sometimes a bottom rail.

Bar Chair A device used to support reinforcing bars during the placing of concrete.

Base Shoe A quarter round trim piece used to cover the joint between the finish flooring and the base.

Batter Boards Horizontal boards offset from the building line and set prior to excavation, used to indicate a specific location, such as the corner of a building.

Bearing Capacity The unit load, in pounds or kips per square foot, that can be safely supported by the soil.

Bed Joint The horizontal mortar joint in masonry work.

Bitumen A substance derived from petroleum or coal used to resist water penetration, such as asphalt or coal-tar pitch.

Blind Nailing Driving nails in such a way that the nail heads are not visible.

Board Foot A unit of measure for lumber equal to the volume of a board $12'' \times 12'' \times 1''$.

Book Matching A way of placing successive veneers sliced from the same flitch so that alternate sheets appear as mirror images.

Brass   An alloy of copper and zinc that is corrosion resistant and very workable.

Bridging   Crossed braces used between joists to stabilize them.

Bronze   An alloy of copper and tin.

BTU   British Thermal Unit: the amount of heat required to raise one pound of water 1 degree Fahrenheit.

Bullnose   The rounding of an exposed edge, such as a tile or wood trim piece.

Butt   A hinge applied to the edge of a door in which only the butt end remains visible.

C

Caisson   A waterproof box-like structure in which construction work can be done below water level.

Camber   A slight upward curve built into a member to compensate for deflection.

Cant Strip   A beveled strip used to avoid a sharp bend in roofing material.

Capillary Action   The tendency of water to move into small spaces, regardless of gravity.

Caulking   The application of a compound to seal the joint between two materials or surfaces.

Cement   A material that is able to unite non-adhesive materials into a solid mass.

Ceramic Veneer   Terra cotta facing applied as a finish wall material.

Chase   A recessed space or passage through a wall or other element to accommodate pipes or ducts.

Cofferdam   A watertight, temporary structure placed under water and pumped dry to allow construction work to be performed.

Cold Joint   A joint formed when a concrete surface hardens before the next batch of concrete is placed against it.

Collar Beam   A horizontal tie beam connecting two opposite rafters at a level above the wall plates.

Combustible   Capable of burning.

Concrete   A mixture of fine and coarse aggregates, portland cement, and water.

Control Joint   A groove in a concrete structure made to predetermine the location of cracks.

Coping   A protective cap of brick, stone, or concrete used at the top of a wall to prevent water penetration.

Cricket   A flashing saddle used on a sloping roof to divert water around a chimney.

Culls   Rejected material whose quality is too low to be used.

Curing   Maintaining concrete at the proper moisture and temperature after it is cast.

Cut and Fill   Earth that is removed (cut) and earth that is added (fill) in grading.

D

Dado   A rectangular grooved wood joint. Also, the lower part of a wall below a molding or other trim.

Dampproofing   The materials and methods used to prevent moisture from penetrating a building at or below grade.

Decibel   A logarithmic measure of sound intensity level.

Double Glazing   Two sheets of glass with an air space between, to insulate against the passage of heat or sound.

Dry Pack   To fill a confined space with a damp concrete mixture, by packing in tightly.

Dry Rot   Timber decay due to fungus, in which pockets of dry powder develop.

Drywall   A masonry wall built without mortar. Also, an interior wall or ceiling of gypsum wall board construction.

E

Earthwork   *See* Grading.

Efflorescence   The whitish powder of crystallization, caused by water soluble salts, which comes to the surface when water evaporates from brick.

Elastomer   A material having the qualities of rubber.

Enamel   A smooth and hard pigmented paint that uses varnish as the vehicle.

Entrained Air   Tiny air bubbles intentionally incorporated into mortar or concrete during mixing to resist freezing action.

Epoxy   A synthetic resin having excellent adhesive properties.

Escutcheon   A metal plate around a knob and/or keyhole of a door.

Excavation   The digging or removal of earth.

Expansion Joint   The complete separation, from top of footing to the roof, of adjacent building parts to allow for expansion and contraction caused by temperature changes.

F

Fire Brick   Brick made from fire clay that has great resistance to high temperatures.

Fireproof   Describing construction that will not burn or is highly resistant to fire.

Firestops   Horizontal blocks placed between studs to resist the spread of fire.

Flame-Spread Rating   A classification of finish materials indicating the rate at which flame will spread.

Flashing   The application of an impervious material to prevent water penetration at joints formed by different materials or surfaces.

Flitch   A large timber from which veneers are cut.

Footing   The part of a foundation that spreads the load over a large area of soil.

Forms   The molds into which concrete is placed until it has hardened.

Foundation   The part of a building's structure that transmits the building's load to the underlying soil.

Furring   Wood or metal strips used to make a plane surface; also a cavity within a wall or ceiling.

G

Galvanic Action   The deteriorating reaction between dissimilar metals that are in contact in the presence of moisture.

Galvanizing   The process of applying a coating of zinc to iron for protection against corrosion.

Glazing   The work of installing glass in a frame.

Glulam   A glued laminated timber used structurally.

Grading   Removing and/or adding earth in order to bring the ground surface to a specified elevation or profile.

Grillage   A framework of horizontal members used to spread a structural load over a larger area.

Grout   A fluid mixture of portland cement, sand, and water used to fill joints and cavities of masonry or tile.

Grubbing   The removal from a site of unwanted roots, stumps, and so forth, during excavation.

Gunite   Pneumatically applied concrete.

Gutter   A trough at the edge of a roof used to carry off rainwater.

Gypsum   Hydrated calcium sulfate, from which gypsum plaster and plasterboard are made.

Gypsum Board   A prefabricated form of plaster used in place of conventional three-coat interior plaster.

H

Hand of Door Convention relating to door swing, used when ordering door hardware.

Header Course A masonry course in which the flat, short ends of the units are exposed.

Hinge A device on which doors, windows, cabinets, and so forth, turn or swing, to open and close.

Hip The exterior edge formed by the meeting of two adjacent sloping roof surfaces.

Hollow Core Describing a door in which veneer faces are glued to an inner skeleton framework.

Honeycomb A void left in concrete that is poorly mixed or placed.

Hospital Tip Rounded top edge on a hinge, designed for cleanliness and to avoid catching garments.

I

Insulating Glass *See* Double Glazing.

Insulation A material that provides high resistant to sound transmission or heat flow.

Intumescent Paint Paint that swells up when exposed to excessive heat, and thus resists flame spread.

J

Jack Rafter A short rafter between hip rafter and eave, or between valley and ridge.

Jalousie A window or door blind made of fixed or movable horizontal slats.

Jetting Placing piles using high pressure water jets.

K

K Value The thermal conductivity of a material.

Kalamein Door A type of fire-resistive door that has a solid wood core covered with sheet metal.

Keene's Cement A quick-setting gypsum plaster used in areas exposed to moisture.

Kerf A cut or notch in a material.

L

Laitance Mortar or grout scum on the surface of concrete.

Latch A beveled fastening device that automatically slides into position when the door is closed.

Ledger Horizontal member supporting joists. Also sometimes called ribbon, girt, purlin, or stringer.

Lewis Bolt A round threaded metal device with a bell-shaped end that is used to anchor stone.

Lock A mechanical fastening device with a rectangular locking bolt that is projected manually or with a key.

M

Mandrel Solid core used in driving a shell pile into the ground. When the mandrel is removed, the shell is filled with concrete.

Mastic A permanently plastic waterproof adhesive material used in sealing joints.

Matrix The binding or cementing material in mortar, concrete, or terrazzo.

Modular Describing a system composed of standardized units or sections used for simplified construction or flexibility.

Monel An alloy of nickel and copper that is resistant to corrosion and used for roofing, kitchen equipment, etc.

Mortar A mixture of cement, lime, sand, and water used to bond bricks or stone in masonry work.

Mullion A vertical member between windows or doors.

Muntin A short, secondary member within a window frame, either vertical or horizontal.

N

Neat   Describing cement or plaster that has been mixed with water only, without sand or lime.

Needle Beam   A short beam passed through a wall to provide temporary support.

O

Oil Canning   The wavy distortions of glass or metal panels, often seen in curtain wall construction.

P

Panic Hardware   A door-latching assembly that will open the door if subjected to pressure.

Parging   The application of plaster to the back of masonry walls.

Particle Board   A manufactured panel of wood particles and binders that is bonded together under heat and pressure.

Perlite   A lightweight volcanic rock used as an aggregate in lightweight concrete or plaster.

Perm   The unit of water vapor transmission, equal to the flow of one grain of water vapor through one square foot of surface per hour under the pressure of one inch of mercury.

Permeability   The property of permitting passage of water vapor through a material without causing rupture or displacement.

Pervious   Permitting leakage or flow of water.

Pigment   The solid component of paint, consisting of finely ground material, which gives the paint its color.

Pile   A vertical member driven into the ground to support the vertical load of the structure above.

Pith   The heart center of a log.

Plaster   A mixture of portland cement (exterior plaster) or gypsum (interior plaster) with sand and water, which is applied in layers, or coats, which harden and dry.

Platform Framing   A method of framing wood stud walls in which the studs are one story in height and the floor joists bear on the top plates of the wall below.

Portland Cement   The cementitious binder used to make concrete and mortar.

Putty Coat   The final smooth coat of interior plaster.

R

R Factor   A measure of thermal resistance.

Raked Joint   A mortar joint that has been cleaned of mortar for about ¾ inch back from the face.

Reglet   A slot in which roofing material or flashing is inserted, generally in a vertical wall surface.

Reverberation   The persistence of sound in an enclosed space after the source has stopped.

Roping   The arrangement of cables used to hoist an elevator.

Rowlock   A masonry course in which the units are laid on edge, with their ends exposed.

S

Sabin   The unit of sound absorption, equivalent to the absorption of one square foot of open window.

Sash   A frame in which the panes of a window or door are placed.

Screed   A strip of material placed at intervals along a wall to be plastered to gauge the thickness of the plaster.

Sealant   Material used to prevent the passage of liquid across a joint or opening.

Shakes   Hand-split wood shingles.

Sheepsfoot Roller   A large-toothed roller used for the compaction of soil.

Sheeting Vertical construction used to temporarily hold the face of an excavation during construction.

Shellac A solution of refined lac resin and denatured alcohol used as a clear coating.

Shoring Temporary support for a portion of a building.

Sleeper A horizontal member used to support a structure above, such as one of the wood strips between a concrete slab and finish wood floor.

Slump Test A test for mixed concrete to determine consistency and workability.

Slurry A soupy mixture of water and clay, or water and portland cement.

Soil A natural material, formed of decomposed and disintegrated parent rock, that can support plant life.

Soil Boring Log A graphic representation of the soils encountered in a test boring.

Sound Transmission Class A rating for the evaluation of a particular construction cross-section in the transmission of airborne sound.

Stile The upright or vertical edge of a door or window.

Stretcher A brick laid with its length parallel to the length of the wall.

Stucco A mixture of sand, portland cement, lime, and water, that may be tinted and applied as an exterior plaster finish.

T

Terne Plate Steel coated with lead and tin used for roofing and flashing.

Terrazzo Flooring material made from small chips of marble set in cement and polished.

Test Boring A hole drilled into the ground, from which samples of undisturbed subsurface soils are obtained for laboratory inspection and testing.

Test Pit An excavation made to expose the subsurface soils for in-place examination.

Tremie A pipe or tube used to deposit concrete under water.

U

U Factor The thermal conductivity of a wall section, expressed in BTUs per hour, per degree Fahrenheit, per square foot.

Urethane A synthetic resin from which paints and insulation foams are manufactured.

V

Valley The interior trough formed by the intersection of two sloping roof surfaces.

Vapor Barrier An airtight layer used to prevent moisture from a warm interior from passing into and condensing within a cold wall or ceiling area.

Vehicle The liquid portion of a paint that holds the pigment or solids in suspension.

Vermiculite A lightweight aggregate used in lightweight concrete or plaster.

Vitrification The fusion of a clay product after firing, which makes it impervious.

W

Warp Distortion of timber during seasoning caused by changing moisture content.

Water-Cement Ratio The ratio of water to cement in a concrete mix, which is the main factor that determines concrete strength.

Waterproof Impervious to water or water vapor, even under pressure.

Water Stop A device used at a joint as a sealant, usually to prevent the passage of water.

Water Table The level below at which the earth is fully saturated with water.

Weatherstripping The means employed to make exterior openings weathertight.

Weep Holes Small openings left in retaining walls, sills, aprons, foundations, etc., which permit water to drain.

Wythe Each individual vertical tier of masonry in a cavity wall.

BIBLIOGRAPHY

The following list of books is provided for candidates who may wish to do further research or study in Building Design & Construction Systems. Most of the books listed below are available in college or technical bookstores, and all would make welcome additions to any architectural bookshelf. In addition to the course material and the volumes listed below, we advise candidates to review regularly the many professional journals, which are available at most architectural offices.

A Dictionary of Building
Scott
Penguin Books

Architectural Graphic Standards
Ramsey and Sleeper
John Wiley and Sons

Basic Building Data
Graf
Van Nostrand Reinhold

Building Construction
Huntington and Mickadeit
John Wiley and Sons

Building Construction Handbook
Merritt
McGraw Hill Book Co.

Building Construction Illustrated
Ching
Van Nostrand Reinhold

Construction Materials
Hornbostel
John Wiley and Sons

Construction Principles, Materials and Methods
Schmitt, Olin, and Lewis
American S&L Institute Press

Dictionary of Architecture and Construction
Harris
McGraw Hill

Fundamentals of Building Construction Materials and Methods
Allen, Edward
John Wiley and Sons

Standards for Access for Persons with Disabilities
ANSI A117.1
American National Standards Institute

Materials and Methods for Contemporary Construction
Hornbostel and Hornung
Prentice Hall

International Building Code
International Code Council

Lesson 1

1. **C** The purpose of a control joint is to create a weakened section, so that shrinkage cracking will occur at the joint, rather than in a random fashion.

2. **A** As hot weather tends to shorten concrete's setting time, one would add a retarding agent to the mixture. This would counteract the weather's effect by lengthening the setting time.

3. **B** See page 16.

4. **A** Lightweight concrete, whether structural or nonstructural, provides good thermal insulation (I) and resistance to fire (III). It has low density (II is incorrect) and relatively high cost (IV is incorrect), and it is easy to handle and place (V is incorrect). The correct combination of answers is found in choice A.

5. **D** The strength of concrete depends on a number of factors, but primarily on the water-cement ratio.

6. **C** After concrete is placed, the proper humidity and temperature must be maintained for some time in order to assure satisfactory hydration of the cement. For this purpose, a number of methods may be employed, including covering the concrete with a membrane (A), supplying additional moisture by spraying (B), and covering the concrete with moist sand (D). Blowing with fans (C) is not normally done, as this would probably result in excessive evaporation of water, which is precisely what curing attempts to avoid.

7. **D** The concrete pour in this question was improper for two reasons: it was dropped from an excessive height, and it was deposited at one point, instead of uniformly along the form. Both of these procedures tend to cause segregation of the mix, and no further treatment will solve this problem. The only solution, therefore, is to remove and discard the concrete.

8. **B** Concrete is strong in compression, but weak in tension. Therefore, reinforcing steel is embedded in the concrete to resist tension, while the concrete itself resists compression. The combination of the two materials is known as reinforced concrete.

9. **D** The maximum size of coarse aggregate permitted is ⅕ of the narrowest dimension between forms, or ⅓ of the thickness of slabs, or ¾ of the minimum space between reinforcing bars. Thus, III and V are correct (choice D).

10. **B** Precast concrete offers a number of advantages, including better quality control (A), all-weather construction (C), and often, greater economy (D). While precast concrete can be steam cured, which provides better quality control, the curing time is not necessarily reduced (B).

Lesson 2

1. **C** See page 39.

2. **A** This is an easy question, as it is apparent that glass block is the only material listed that is able to transmit light.

3. **C** Face brick is more uniform in size and color than building brick. Its durability and weather resistance are the same as building brick of the same grade (A and B are incorrect). Face brick is only available in certain sizes, textures, and colors (D is incorrect).

4. **A** Concrete block is usually less expensive per square foot of wall than brick (I). Concrete blocks are also much faster to erect than brick because of their larger size: one concrete block has the same volume as

8 or 12 bricks (IV). Because they are larger and heavier, concrete blocks are harder to handle (II is incorrect). Blocks are not easier to reinforce (III is incorrect), and in general, they are less attractive than brick (V is incorrect).

5. **B**   Efflorescence is a whitish powder that sometimes comes to the surface when water evaporates from brick or other masonry. Choices A, C, and D all help to keep water from penetrating the masonry and therefore tend to reduce efflorescence. Wetting the masonry units (B) may be desirable, particularly for brick, but is does nothing to reduce efflorescence.

6. **D**   Hydrated lime is used in a mortar mixture to improve workability and water retentivity, although it reduces the strength of the mortar (A and B are incorrect, D is correct). C is also incorrect.

7. **B**   The variations in the color and hardness of brick are caused by the varying characteristics of the clays from which they are made (II) and the temperatures at which they are fired (IV).

8. **D**   See Figure 2.4.

9. **D**   Water tends to collect and remain in troweled joints, so it is likely that if left alone, the mortar would eventually deteriorate, causing leaks and loose units. It is important that joints be well-filled, tightly compacted, and of the proper shape. Troweled joints are a poor choice on an exterior wall, for they collect water, while weather-struck joints shed water. Adding mortar to an existing joint (C) is always poor practice, so the only alternative is to redo the wall.

10. **D**   None of the first three choices provides the weather resistance of a proper mortar joint. Plastering the wall is a reasonable

alternative, but it changes the architectural character and may not be acceptable to the owner and/or architect. Rebuilding the wall is still a better choice.

Lesson 3

1. **B**   Heavy timber construction consists of wood beams, girders, and columns that are very large, as specified by the building code. Such construction is fire resistant because massive wood members burn very slowly.

2. **C**   Insects, decay, and fire are continual risks to wood construction. Keeping underfloor wood members dry and well-ventilated (I) reduces the risk of damage from insects or decay. Intermittent wetting provides a favorable environment for decay, but complete submersion in water (II), so that air is excluded, minimizes the possibility of decay. Decay-resistant species of wood, such as redwood, are often used for members close to the earth, but these are generally softwood, not hardwood (III is incorrect). Surface coatings are usually ineffective in preventing termite infestation (IV is incorrect). V is correct: the use of intumescent paint reduces the rate of flame spread.

3. **A**   Wood trusses, laminated beams, and girders are all possible substitutes for conventional joists, but their use would be impractical and far more expensive than providing additional joists.

4. **C**   Nails and other metal fasteners were not massed produced until the 19th century, which completely changed all types of wood joinery.

5. **A**   See page 56.

6. **D**   Wood with a moisture content of 30 percent is considered unseasoned, or green, and will shrink if used in construction.

7. **A**   See page 60.

8. **C**   See page 60.

9. **B**   Plainsawed lumber has a more distinctive grain pattern than quartersawed or rift cut. Factory and shop lumber refers to lumber used for remanufacture, not to grain pattern.

10. **D**   All of the listed factors affect the strength of a wood member.

Lesson 4

1. **C**   Cast iron has a high carbon content and is strong in compression, but weak in tension.

2. **A**   Among the metals listed, bronze is most resistant to galvanic deterioration. Refer to the list on page 78.

3. **D**   Steel is the most widely used metal in building construction, including those items classified as miscellaneous metalwork.

4. **C**   Open web steel joists are shop-fabricated (III), small, lightweight, standardized steel trusses (I and II). They are strong in the vertical direction, but weak in the flat, horizontal direction (IV is incorrect).

5. **B**   Steel is strong and relatively inexpensive (I and III). However, it is susceptible to corrosion and lacks fire resistance (IV and V). Steel usually requires a great deal of precise shop fabrication (II is incorrect).

6. **A**   Aluminum has all the properties listed, except high resistance to galvanic action. See pages 80–81.

7. **D**   Lightweight metal framing systems have all of the advantages listed, except fire resistance. Although metal is incombustible, it loses so much strength when exposed to a severe fire that it may fail or deform excessively.

8. **B**   Galvanizing iron and steel protects it against deterioration from corrosion.

9. **B**   Metal decking can support live and dead loads, can serve as permanent formwork, and is usually corrugated or ribbed (I, IV, and V are correct). Decking is usually welded to the supporting members (II is incorrect). Whether plain or galvanized, metal decking must have a covering of concrete for permanent weather protection (III is incorrect).

10. **D**   An advantage to using aluminum is that complex sections, such as those used for window assemblies, can easily be produced by extrusion. In this process, a heated billet of aluminum is squeezed through a die to produce the desired shape in any length.

Lesson 5

1. **C**   Tile roofs are heavy, permanent, and fireproof (I, III, and IV). However, they are expensive (II is incorrect), and they are only suitable for relatively steep roof pitches (V is incorrect).

2. **B**   Slabs on grade are often placed over a granular fill several inches thick in order to prevent groundwater from being drawn into the slab by capillary action.

3. **C**   Loose fill is an inappropriate insulation because the fill would tend to slide along the sloping surface and accumulate at the low end. Such insulation should only be used in flat air spaces, such as attics, or in completely filled wall cavities.

4. **A**   Vapor barriers should always be installed on the warm or room side of construction, in this case just above the finish ceiling. Vapor barriers are never installed between rafters, because the

sheet must be continuous to be effective in preventing the passage of water vapor.

5. **C** Expansion joints cannot reduce, prevent, or cause movement due to temperature change. Their purpose is to provide a complete separation, which will permit movement, while maintaining structural integrity and weathertightness.

6. **A** See page 94.

7. **D** Condensation can be controlled by using insulating glass (I), applying a vapor barrier (II), and providing ventilation (V). Attic insulation (III) is used for thermal control, not condensation control. And perimeter drain tiles (IV) are used to carry groundwater away from a structure.

8. **B** See Figure 5.4.

9. **B** Flashing is used to provide a seal and prevent water penetration at joints and intersections, such as those in choices A, C, and D. The standing seams of a sloping metal roof form a seal and therefore do not require flashing (B).

10. **D** Referring to Table 5.1, one can see that the R value for a one-inch plastic panel is 6.00, which is greater per inch of thickness than any other type shown.

Lesson 6

1. **B** Patterned glass obscures vision and reduces light transmission.

2. **C** Skylight glazing must be safe from breaking and falling, which might endanger occupants. Laminated, tempered, or wired glass are appropriate to specify for skylights.

3. **D** A tactile finish is used on operating hardware on doors leading to areas that might be dangerous to blind people.

4. **C** Aluminum door and window sections are strong and durable (B and D are incorrect), and have relatively low initial expense (A is incorrect). However, special precautions must be taken to prevent galvanic action (C).

5. **B** Referring to Table 6.1, the glazing permitted in a Class B labeled fire door is 100 square inches. Dividing this area by the four-inch width of glazing results in a maximum allowable height of 25 inches.

6. **B** Revolving doors can carry a continuous two-way flow of pedestrian traffic (IV is incorrect, V is correct). Very little cold (or warm) outside air is permitted to enter the building (III is correct). However, revolving doors do not provide for emergency exiting or handicapped access (I and II are incorrect).

7. **A** See page 114.

8. **A** Plastic substitutes for glass are available in transparent sheets (C is incorrect), are fire-resistant (B is incorrect), and are usually more expensive (D is incorrect). However, plastic is generally less durable than glass (A is correct).

9. **C** Panic hardware is used on exit doors to quickly and easily open the doors in case of emergency. Knurled knobs (A) are used as a tactile finish to warn blind people that the doors lead to a dangerous area. A lever handle (B) is a horizontal handle used to open a door and is not suitable for emergency exiting. A fusible link (D) melts in case of fire, thereby closing a door or damper to prevent the spread of fire.

10. **D** Wood windows are relatively inexpensive, durable, and widely available (II, IV, and V). However, they require periodic maintenance and are not fire-resistive (I and III are incorrect).

Lesson 7

1. **C** Asphalt tile, terrazzo, and elastomeric-type flooring lack resilience, while cork flooring is highly resilient.

2. **D** Vermiculite is an expanded mica used as an insulating fill or as aggregate in a plaster mix to make it fire-resistant and lightweight.

3. **B** Statements I, II, and V are advantages of using a suspended acoustical tile ceiling system. Statements III and IV are incorrect: acoustic tiles are fragile and provide no fire rating.

4. **D** A plumb surface is created by attaching metal furring channels to the concrete walls, over which metal lath and plaster are applied. Applying lath directly to rough concrete would not result in a smooth, plumb surface. Also, because of potential moisture penetration, using gypsum or plywood would be risky.

5. **B** Gypsum board, or drywall construction, has many advantages over wet plaster, but greater durability is not one of them. Usually, wet plaster applied in three coats is thicker, more dense, and more durable.

6. **C** The most important factor in painting work is surface preparation. The surface must be clean and dry if the paint is to perform satisfactorily.

7. **C** Ceramic tiles are permanent, waterproof, durable, and easily maintained (A, B, and D are incorrect). However, they are more costly than Keene's cement plaster.

8. **A** The scratch coat is the first plaster coat (I), and its ingredients are correctly described in statement II. It is generally applied over gypsum lath or metal lath (III is incorrect), it may be applied by hand or machine (IV is incorrect), and it is about ¼ inch thick (V is incorrect).

9. **A** Wood flooring is durable, comfortable, attractive, and easy to maintain. However, all wood is susceptible to swelling and shrinkage caused by changes in humidity (A), and therefore, sufficient expansion space must be provided at the floor's perimeter. Wood flooring generally has a hard, dense surface to withstand heavy wear (B is incorrect). Wood floors tend to burn very slowly in case of fire (C is incorrect), and some types of wood flooring are very resistant to the effects of oil, grease, and mild chemicals (D is incorrect).

10. **B** Floor finishes, such as urethane, are applied directly to concrete, wood, or metal to produce a hard, durable, wear-resistant, easily maintained, greaseproof floor finish.

Lesson 8

1. **D** All of the above. Building codes are intended to protect the public health, safety, and welfare.

2. **A** The two model building codes published in the United States are the International Building Code and the National Fire Protection Association (NFPA) 5000. The other three model building codes, Uniform, National, and Standard were discontinued when the International Building Code was published.

3. **B** Business occupancies are considered the least dangerous because the occupants should be familiar with their surroundings and are expected to be awake while in the building. Assembly occupancies have an increased potential of panic; residential occupancies have an increased potential for problems due to people's sleeping in the facility. There is no commercial occupancy.

4. **A** Type II N is considered noncombustible construction. The rest are all considered combustible.

5. **D** Building area is determined by the occupancy and type of construction with the resultant area increased based on percentage of building frontage accessible to a public way or open space, whether or not the building has an automatic sprinkler system, and the number of stories of the building.

6. **A** Wide flange structural steel members with fireproofing can be fire rated relating to potential collapse for one, two, three, or four hours. Exterior curtain walls are not capable of withstanding fire in any application. Enclosed vertical exit stairs are only rated as one hour or two hours. Doors are fire rated for 20 minutes, 30 minutes, 45 minutes, one hour, 90 minutes, two hours, or three hours.

7. **B** The Steiner Tunnel Test, also known as ASTM E 84, UL 992, and NFPA 255, is used to determine the flame spread rating of finishes for interior walls and ceilings.

8. **C** The means of egress consists of three elements: the exit access, the exit, and the exit discharge. Exit doors could be included in any of the three. Corridors are part of the exit access with comparable elements being the exit passageways for the exit, and exit courts for the exit discharge.

9. **B** Three exits are required where there are more than 500 occupants, and four exits are required where there are more than 1,000 occupants. Other areas typically require only two exits, but where the path to the first exit exceeds the allowable travel distance, a third exit would be required.

10. **D** The ICC/ANSI A117.1 American National Standard "Accessible and Usable Buildings and Facilities," the Fair Housing Accessibility Guidelines (FHAG), the Americans with Disabilities Act Accessibility Guidelines (ADAAG), the Uniform Federal Accessibility Standards (UFAS), and the International Building Code, Chapter 11, all provide information for designing facilities accessible to persons with disabilities.

Lesson 9

1. **D** All of the above. The holistic approach to sustainably designed projects encourages the design team to examine the impact of environmental, economic, mechanical, and aesthetic architectural decisions.

2. **D** None of the above.

Choice I is not correct. The zone of the earth that supports human life (five miles into the earth's crust and five miles into the atmosphere) is an extremely fragile ecosystem. This biosphere that has evolved over millions of years has been dramatically affected by the growth of human activity in the last 150 years.

Choice II is not correct either. While innovative technologies are improving energy efficiency of some building systems, the vast majority of the built environment is energy inefficient.

Choice III is also not correct. Toxic substances have the tendency to expand and affect large areas. For example, the air above the Great Lakes contains evidence of DDT, a toxic pesticide banned in the United States decades ago. It was discovered that DDT is captured in the jet stream, bringing toxic materials from far-away continents, which still use toxic pesticides.

Choice IV is not correct. While recycling is helpful, it is just the beginning of the sustainable design process. The principles of sustainable design say that we need to have more building products that can be recycled and are biodegradable to create a more sustainable ecosystem.

3. C   I, III, and IV

Choice I is correct. Designing with native landscaping is preferred to using exotic or imported plant types. Indigenous plants tend to survive longer, use less water, and cost less.

Choice II is not correct. Placing any structure in a floodplain, even those that resist floodwater, is not desirable. Placing buildings in a floodplain can increase flooding further down stream.

Choice III is also correct. Buildings sensitive to the benefits of solar orientation and passive and active solar gain techniques save energy and are more visually aligned with local climatic conditions.

Choice IV is correct as well. In-fill development and proximity to a variety of transportation options are design principles that benefit the inhabitants and their environment.

4. C   III and IV

I is not correct. Communities that are only residential are not encouraged. Mixed-use development (combining housing, retail, open space, and commercial) is a preferred sustainable design.

II is not correct. Open space should not be designed only for recreation and wild life habitat. Additional uses, such as environmental education, storm water retention, flood control, wetlands drainage, and so on, should be considered in sustainable planning.

III is correct. The Ahwahnee principles support a wide range of interconnected transportation to encourage many options for travel.

IV is also correct. Development that permits opportunities for a diverse number of jobs is a key goal of the Ahwahnee principles.

5. D   All of the above

I is correct. While first cost is not the primary concern of life cycle costing, it is one of the economic factors considered.

II is also correct. The cost of maintenance is part of the evaluation.

III is correct as well. The durability of a product or system is considered in the cost of repair and part of the overall evaluation.

IV is correct because the comparison of product or system life is one of the factors evaluated in life cycle costing.

6. B   I, II, and III

I is correct. LEED has several options for improving Indoor Air Quality (IAQ) including filtering the air system and installing low Volatile Organic Compound (VOC) paints and caulking.

II is also correct. Methods to store, recirculate, and locally distribute rainwater are encouraged.

III is correct as well. Innovative solutions to energy conservation, such as fuel cells, photovoltaic panels, and gas turbine energy production are encouraged in the LEED accreditation system.

IV is incorrect. Unfortunately, the LEED system awards no points awards for designs with strong aesthetics.

7. D   All of the above

All of these consultants (wetlands engineer, energy commissioner, landscape architect, and energy modeling engineer) might be necessary for the holistic approach to sustainable design. The landscape architect should have experience with local, native plant design.

8. **C**   I, II, and IV

I is correct. Computer programs that allow energy modeling of design options allow the architect a quick method of evaluating numerous different solutions.

II is also correct. It is extremely important that the client be able to understand the value of sustainable design solutions.

III is not correct. Art selection is at the client's discretion.

IV is correct. Locating the most energy efficient appliances, plumbing fixtures, and office equipment will improve the energy efficiency of the entire project.

9. **B**   I, III, and IV

I is correct. Solar orientation can affect many architectural design elements, including massing, landscaping, fenestration, and building skin design.

II is not correct. Landscape design should be functional as well as visually pleasing. Landscape design for purely visual impact is not consistent with the sustainable design approach.

III is also correct. Architectural design that understands the context (scale, color, style, texture) of adjacent structures is sympathetic to the sustainable design philosophy.

IV is correct as well. Understanding all site conditions and their potential to assist building's energy systems is helpful. For example, groundwater connected to a heat pump is a good source of supplemental energy for cooling and heating a building.

10. **A**   I, II, and IV

I is correct. Solar shading, whether from landscaping or architectural elements, can regulate the insulation to increase winter light and reduce warm summer sunlight.

II is also correct. Urban heat island effect is the tendency of a building roof to absorb solar radiation during the day and then emit heat radiation during the evening. Roof systems with grass or light-colored roofing material reduce the urban heat island effect.

III is not correct. Sustainable design encourages approaches that reduce the area allocated to parking.

IV is correct. The type, location, and size of building fenestration are key aspects of architectural design for sustainable projects.

EXAMINATION DIRECTIONS

The examination on the following pages should be taken when you have completed your study of all the lessons in this course. It is designed to simulate the Building Design & Construction Systems division of the Architect Registration Examination. Many questions are intentionally difficult in order to reflect the pattern of questions you may expect to encounter on the actual examination.

You will also notice that the subject matter for several questions has not been covered in the course material. This situation is inevitable and, thus, should provide you with practice in making an educated guess. Other questions may appear ambiguous, trivial, or simply unfair. This too, unfortunately, reflects the actual experience of the exam and should prepare you for the worst you may encounter.

Answers and complete explanations will be found on the pages following the examination, to permit self-grading. **Do not look at these answers until you have completed the entire exam.** Once the examination is completed and graded, your weaknesses will be revealed, and you are urged to do further study in those areas.

Please observe the following directions:

1. The examination is closed book; please do not use any reference material.

2. Allow about one hour to answer all questions. Time is definitely a factor to be seriously considered.

3. Read all questions *carefully* and mark the appropriate answer on the answer sheet provided.

4. Answer all questions, even if you must guess. Do not leave any questions unanswered.

5. If time allows, review your answers, but do not arbitrarily change any answer.

6. Turn to the answers only after you have completed the entire examination.

GOOD LUCK!

EXAMINATION ANSWER SHEET

Directions: Read each question and its lettered answers. When you have decided which answer is correct, blacken the corresponding space on this sheet. After completing the exam, you may grade yourself; complete answers and explanations will be found on the pages following the examination.

1. Ⓐ Ⓑ Ⓒ Ⓓ
2. Ⓐ Ⓑ Ⓒ Ⓓ
3. Ⓐ Ⓑ Ⓒ Ⓓ
4. Ⓐ Ⓑ Ⓒ Ⓓ
5. Ⓐ Ⓑ Ⓒ Ⓓ
6. Ⓐ Ⓑ Ⓒ Ⓓ
7. Ⓐ Ⓑ Ⓒ Ⓓ
8. Ⓐ Ⓑ Ⓒ Ⓓ
9. Ⓐ Ⓑ Ⓒ Ⓓ
10. Ⓐ Ⓑ Ⓒ Ⓓ
11. Ⓐ Ⓑ Ⓒ Ⓓ
12. Ⓐ Ⓑ Ⓒ Ⓓ
13. Ⓐ Ⓑ Ⓒ Ⓓ
14. Ⓐ Ⓑ Ⓒ Ⓓ
15. Ⓐ Ⓑ Ⓒ Ⓓ
16. Ⓐ Ⓑ Ⓒ Ⓓ
17. Ⓐ Ⓑ Ⓒ Ⓓ
18. Ⓐ Ⓑ Ⓒ Ⓓ
19. Ⓐ Ⓑ Ⓒ Ⓓ
20. Ⓐ Ⓑ Ⓒ Ⓓ
21. Ⓐ Ⓑ Ⓒ Ⓓ
22. Ⓐ Ⓑ Ⓒ Ⓓ
23. Ⓐ Ⓑ Ⓒ Ⓓ
24. Ⓐ Ⓑ Ⓒ Ⓓ

25. Ⓐ Ⓑ Ⓒ Ⓓ
26. Ⓐ Ⓑ Ⓒ Ⓓ
27. Ⓐ Ⓑ Ⓒ Ⓓ
28. Ⓐ Ⓑ Ⓒ Ⓓ
29. Ⓐ Ⓑ Ⓒ Ⓓ
30. Ⓐ Ⓑ Ⓒ Ⓓ
31. Ⓐ Ⓑ Ⓒ Ⓓ
32. Ⓐ Ⓑ Ⓒ Ⓓ
33. Ⓐ Ⓑ Ⓒ Ⓓ
34. Ⓐ Ⓑ Ⓒ Ⓓ
35. Ⓐ Ⓑ Ⓒ Ⓓ
36. Ⓐ Ⓑ Ⓒ Ⓓ
37. Ⓐ Ⓑ Ⓒ Ⓓ
38. Ⓐ Ⓑ Ⓒ Ⓓ
39. Ⓐ Ⓑ Ⓒ Ⓓ
40. Ⓐ Ⓑ Ⓒ Ⓓ
41. Ⓐ Ⓑ Ⓒ Ⓓ
42. Ⓐ Ⓑ Ⓒ Ⓓ
43. Ⓐ Ⓑ Ⓒ Ⓓ
44. Ⓐ Ⓑ Ⓒ Ⓓ
45. Ⓐ Ⓑ Ⓒ Ⓓ
46. Ⓐ Ⓑ Ⓒ Ⓓ
47. Ⓐ Ⓑ Ⓒ Ⓓ

1. Which type of glass is appropriate to use in areas exposed to fire hazards?

   A. Heat-absorbing glass

   B. Heat-strengthened glass

   C. Wired glass

   D. Tempered glass

2. Masonry joints that lack sufficient mortar are often repaired by adding fresh mortar while the surrounding mortar is still green. This process is called

   A. raking.

   B. pointing.

   C. troweling.

   D. jointing.

3. Spandrel glazing in a curtain wall system should be

   A. double-strength glass.

   B. tempered glass.

   C. heat-absorbing glass.

   D. heat-strengthened glass.

4. Life-cycle costing is an economic evaluation of architectural elements that include which of the following factors?

   I.   First cost

   II. Maintenance and operational costs

   III. Repair costs

   IV. Replacement cost

   A. I

   B. II, III, and IV

   C. II and IV

   D. All of the above

5. In laying a conventional wood strip tongue-and-groove floor, the strips are usually attached to the subfloor by blind nailing. With regard to the last strip, closest to the wall, it is

   A. attached with waterproof mastic.

   B. attached by face nailing.

   C. also attached by blind nailing.

   D. not attached, but left loose for expansion.

6. The slump test is used to measure the

   A. compressive strength of concrete.

   B. tensile strength of concrete.

   C. soundness of concrete aggregates.

   D. workability of concrete.

7. Structural steel members are to be used in a structure where they will be exposed to severe weather conditions. The members have been cleaned and prepared at the mill, where a priming coat of red-lead alkyd-based primer has been applied. After they are in place, you should specify a finish coat of

   A. coal-tar enamel.

   B. alkyd enamel.

   C. oleoresinous paint.

   D. any of the above.

8. With reference to the detail shown below, which of the following statements is most correct?

   A. The foundation drain is placed in gravel fill in order to minimize the cracking or breakage of individual drain tiles.

   B. The waterproof rigid insulation between the concrete slab and foundation wall may be omitted in areas where the minimum daytime winter air temperature remains above freezing.

   C. The granular fill beneath the slab may consist of crushed stone, coarse slag, or gravel.

   D. Even with the waterproof membrane placed as shown, floor coverings may be damaged by the rise of capillary moisture from the ground.

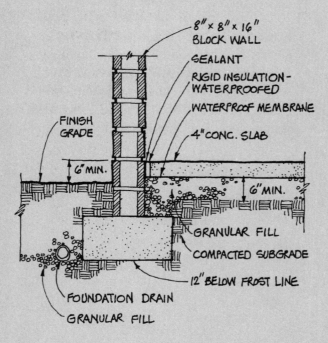

9. Referring to the drawing of the previous question, what is the most obvious error in the detail, considering conventional practices?

   A. The foundation drain should be located just below the bottom of the footing.

   B. The finish grade should slope away from the foundation wall.

   C. The waterproof membrane should be located below the granular fill.

   D. The finish grade should be set at least 12 inches below the finish slab level.

10. Prior to the 18th century, iron was used only sparingly, for fastenings or in decorative ways. All of the reasons below account for this limited use *EXCEPT*

    A. it offered poor resistance to weather exposure.

    B. it lacked classical precedents and therefore was unpopular.

    C. it could not be produced in large sizes or in great quantity.

    D. iron ore was scarce and difficult to mine.

11. Concrete made from Type IV low-heat portland cement would be most appropriately used

    A. in massive concrete pours.

    B. for general overall use.

    C. during very cold weather.

    D. in high alkaline areas.

12. Bulb tees are generally used

    A. in foundation work involving caissons.

    B. in underpinning as a form of temporary support.

    C. in gypsum concrete roof deck construction.

    D. as water stops in below-grade concreting.

13. At the conclusion of a workday, a masonry wall remains unfinished. As the weather report calls for a rainstorm that evening, you should advise the contractor to

A. cover the exposed top of the unfinished wall with loose boards or roofing paper.

B. protect the exposed top with a positive waterproof cover, securely weighted down and hung over each side of the wall at least two feet.

C. ignore the rain, unless the temperature drops below 40 degrees.

D. ignore the rain, as masonry, whether finished or not, cannot be harmed by water.

14. Combustible types of construction include all but which of the following?

A. Type II N

B. Type III N

C. Type V N

D. Type V one hour

15. The planning phase of a sustainably designed architectural project should include which of the following elements?

I. Native landscaping that is aesthetically pleasing and functional

II. Designing structures in the floodplain that can resist the forces of flood waters

III. Consideration of sun orientation, topographic relief, and the scale of adjacent buildings

IV. Locating projects within existing neighborhoods that are adjacent to public transportation

A. I and II

B. I and III

C. I, III, and IV

D. All of the above

16. The need for three exits can be based on which of the following?

I. Occupancy type

II. Type of construction

III. Number of occupants (more than 500 but not more than 1,000)

IV. Travel distance to an exit

V. Intervening rooms

A. I and II C. III and V

B. III and IV D. IV and V

17. Which of the following statements concerning fire doors is NOT true?

A. Class A fire doors are rated for 3 hours and are not permitted to have windows in them.

B. A 1½-hour-rated fire door may be either Class B or Class D, and is permitted to have up to 100 square inches of wired glass per leaf.

C. Class E doors are rated ¾-hour fire resistance in exterior walls with up to 720 square inches of wired glass.

D. Carpeting is not permitted to pass under any fire door, regardless of the rating of the door, the flame spread rating of the carpet, or whether the door is arranged to be held open by a magnetic release device to permit self-closing in the event of fire.

18. An incombustible building material is one that

A. has at least a one-hour fire rating.

B. is not made of wood, fabric, or paper.

C. will not burn.

D. will smolder but not support flames.

19. With reference to the detail shown below, the cant strip is indicated as a continuous filler piece cut on a 45-degree angle. The purpose of this angle is to

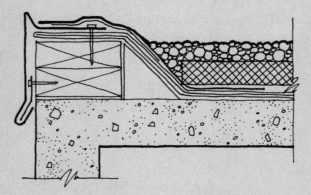

A. provide a moderate bend in the roofing sheets and thereby avoid potential damage to the roofing.

B. assure that roofing water will run off the gravel stop and flow away from the roof edges, toward the drains.

C. avoid 90-degree angle creases or pockets in which water may collect and possibly freeze in the winter.

D. conserve material, as two cant strips may be cut from a single rectangular block.

20. Sustainably designed architecture requires attention to which of the following building elements?

I.   Solar shading devices

II. Urban heat island effect

III. Increased parking

IV. Fenestration and glazing

A. I, II, and IV

B. I and IV

C. I and II

D. All of the above

21. Among the following statements concerning glued laminated timber, which is true?

A. Laminated members are manufactured from the longest pieces of lumber commonly available.

B. Laminations may vary in thickness, but in no case are they ever less than 1½ inches thick.

C. Glued laminated timbers in straight lengths can readily span 60 feet or more.

D. The classification for the finest appearance of glued laminated members is "architectural" grade.

22. According to the ANSI handicapped standards, objects projecting from walls, such as public telephones, which are mounted between 27 and 80 inches above the finished floor, may not protrude more than 4 inches into passageways. Objects lower than 27 inches, such as trash receptacles, may protrude

A. an equal amount.

B. no amount at all.

C. a variable amount.

D. any amount.

23. An abundance of native stone was used in ancient Egyptian architecture because it

A. was relatively easy to handle.

B. was low in cost, relative to other materials.

C. was the most stable and permanent material available.

D. had a high insulating value.

24. Which of the following statements concerning the water-cement ratio of a concrete mix is NOT true?

   A. The water-cement ratio is based on volume, not weight.

   B. A low water-cement ratio reduces the shrinkage of concrete and increases its durability.

   C. The strength of concrete remains the same for a given water-cement ratio, irrespective of the amount of aggregate.

   D. Provided there is sufficient water to hydrate the cement and to avoid air being trapped because the mix is too stiff to compact thoroughly, the strength of concrete increases as the water-cement ratio increases.

25. Building codes are intended to protect which of the following?

   A. Public health

   B. Public welfare

   C. Public safety

   D. All of the above

26. Aluminum window sections are frequently coated with methacrylate lacquers or strippable plastic coatings prior to installation in a rough-framed structure. The purpose of these temporary coatings is to

   A. minimize corrosive damage from galvanic action.

   B. minimize corrosive damage from alkalies.

   C. reduce deterioration caused by normal weathering.

   D. keep the sections clean until construction is completed.

27. Clay roofing tiles are generally attached by means of

   A. steel or hardened steel nails.

   B. annular or helical nails.

   C. copper or aluminum roofing nails.

   D. elastic cement or mastic.

28. In which of the situations described below would you most likely employ a needle beam?

   A. To support the spandrel section in a conventional metal curtain wall

   B. To support the load above an opening cut into an existing masonry wall

   C. To support vertical plywood forming sections for poured-in-place concrete walls

   D. To support cantilevered roof loads in exposed plank-and-beam construction

29. Gypsum partition tiles, otherwise known as plaster blocks, are frequently used for lightweight, fire-resistant interior partitions. Such units are generally held together with

   A. gypsum mortar.

   B. portland cement mortar.

   C. waterproof mortar.

   D. type A or B mortar.

30. At the intersection of a wall and ceiling, one might specify a wooden trim shape known as a

   A. crown mold.   **C.** corner bead.

   B. casing mold.   **D.** chair rail.

31. If an Underwriters' Laboratories 1½-hour B label were attached to a door, you would know that

  A. the door required an automatic closing device.

  B. glazed openings would not be allowed in the door.

  C. Both of the above statements are true.

  D. Neither of the above statements is true.

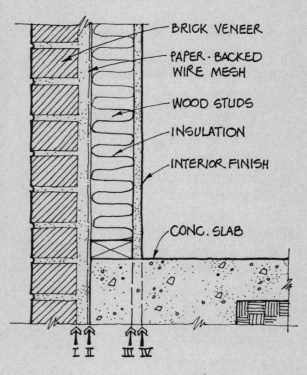

BRICK VENEER

PAPER - BACKED WIRE MESH

WOOD STUDS

INSULATION

INTERIOR FINISH

CONC. SLAB

I II III IV

32. Along which line in the diagram above should a vapor barrier be located?

  A. II only   **C.** I or II

  B. III only   **D.** III or IV

33. LEED, the name of a program that environmentally evaluates sustainable projects, is a checklist that is concerned with which of the following?

  I. Indoor air quality

  II. Storm water

  III. Innovative energy systems

  IV. Aesthetic design

  A. I

  B. I, II, and III

  C. II and III

  D. All of the above

34. Metal casement windows, regardless of the manufacturer, are generally available in identical stock sizes. The standard dimensions of these windows are based on

  A. standard metric measurements.

  B. a standard four-inch module.

  C. standard brick masonry modules.

  D. traditional English measurements.

35. If you specified the use of an intumescent paint, you would expect the coating to

  A. inhibit rust on ferrous metal surfaces.

  B. provide dampproofing on concrete or masonry above grade.

  C. resist abrasion and impact when used on floor surfaces.

  D. retard the spread of fire and the effects of intense heat.

36. The apartment building on the Rue Franklin in Paris, designed by Auguste Perret, is famous as the first such building to

    A. eliminate all decorative ornament.

    B. employ reinforced concrete framing.

    C. reach great heights using load-bearing masonry walls.

    D. make use of extensive glass areas.

37. Shown below is a section through a terrazzo floor installation. The purpose of the neoprene filler is to

    A. allow for expansion, contraction, or structural movement.

    B. provide a waterproof joint at a point of stress in the floor.

    C. provide a durable accent strip of contrasting color and texture.

    D. provide a preformed cold joint for pouring a large area of terrazzo.

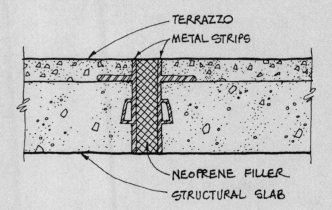

38. In built-up composition roofing systems, one of the principal causes of roofing failure is the presence of blisters, which may be caused by

    I. air between the roofing felts.

    II. moisture in the roofing insulation.

    III. moisture in the roof deck.

    IV. air between the roofing felt and the roof deck.

    A. I only   **C.** I, III, and IV

    B. II and III   **D.** I, II, III, and IV

39. What are the actual dimensions of a nominal $8'' \times 8'' \times 16''$ concrete masonry unit?

    A. $8'' \times 8'' \times 15\frac{5}{8}''$

    B. $8'' \times 7\frac{5}{8}'' \times 15\frac{5}{8}''$

    C. $7\frac{5}{8}'' \times 7\frac{5}{8}'' \times 15\frac{5}{8}''$

    D. $8'' \times 8'' \times 16''$

40. Which of the following paints would you select if low-flame-spread rating were the most important consideration?

    A. Acrylic latex

    B. Polyester-epoxy

    C. Alkyd

    D. Polyurethane

41. Which of the following features apply to the Crystal Palace in London, built in 1851?

    I. It was the first building to employ iron framing.

    II. It was the first building to exhibit great spans.

    III. It was standardized and prefabricated.

    IV. It was, in 1851, the largest building in the world.

    V. It was completely fireproof.

    A. I and II   **C.** I, II, and V

    B. III and IV   **D.** III, IV, and V

42. Select the correct statements.

    I. The principal factors affecting the strength of concrete are the water-cement ratio and the extent of hydration.

    II. Air-entrained concrete is more resistant to freezing and thawing than non-air-entrained concrete.

    III. Concrete having a low water-cement ratio is stronger and more resistant to freezing and thawing than concrete with a high water-cement ratio.

    IV. The length of time during which concrete is moist-cured affects its strength and watertightness.

    V. Excessive bleeding tends to make the surface of concrete weak.

    A. I, III, and V   **C.** II, IV, and V

    B. I, II, and IV   **D.** I, II, III, IV, and V

43. A section through steel floor decking is shown above. What is the purpose of the deformations in the deck?

    A. They simplify the fabrication of the deck.

    B. They improve the safety of the decking during construction by making the surface non-skid.

    C. They lock the concrete slab and decking together to achieve composite action.

    D. They increase the section modulus of the decking.

44. The clear space between a handrail or grab bar used by the handicapped and the wall on which it is installed should be no less than

    A. 1½ inches.

    B. 3 inches.

    C. 6 inches.

    D. 10 percent of its height from the floor.

45. A school district has asked you to reduce the heat loss for a 35-year-old classroom building. The construction of the exterior walls is eight-inch hollow concrete block with no interior finish. Natural light consists of 3-foot by 5-foot wood sash window frames with single glazing in the masonry walls, and these constitute approximately 15 percent of the exterior wall surface. Without preparing a detailed study, you suggest that the most important first step is to

A. apply rigid insulation and gypsum board to the interior surface of the exterior walls.

B. replace all single glazing with double glazing.

C. caulk around all the windows and doors.

D. re-grout the masonry walls.

46. A new building is being constructed adjacent to an existing building whose footings are shallower than those planned for the new structure. The foundation for the new building should be constructed

A. in the usual way, as the existing building's footings will be minimally disrupted.

B. to the same depth as the footings in the existing building.

C. after the footings of the existing building have been extended down to the depth of the footings of the proposed building.

D. in two steps: first on the three non-adjacent sides and then on the remaining side.

47. In the detail of an aluminum skylight frame illustrated here, an arrow is shown pointing to an L-shaped extrusion. The purpose of this extrusion is to

A. strengthen the bending resistance of the overall section.

B. provide a means for removal of condensate.

C. provide an integral ledge on which sun-control panels may be supported.

D. provide a backup safety ledge capable of supporting the glass panel above.

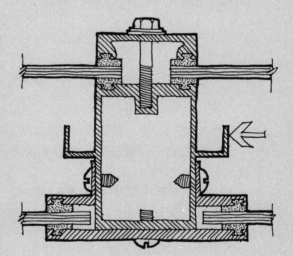

The examination answers and explanations will be found on the following pages.

Do not look at the answers until you have completed the exam.

EXAMINATION ANSWERS

1. **C**   Wired glass is sheet or plate glass into which a thin wire mesh has been embedded. It has great resistance to impact and resists shattering when exposed to high heat. Therefore, it is used in fire doors, exterior walls, vertical shafts, and areas leading to fire escapes.

2. **B**   Because the lack of a full bed or head of mortar can cause a masonry joint to leak, incomplete joints must be repaired while the mortar is still fresh. This process is called pointing. When similar repairs are made to old work, the process is referred to as tuck pointing.

3. **D**   Heat-strengthened glass is plate or patterned glass to which a colored ceramic glaze has been fused. It is strong, opaque, and used almost exclusively in panels for spandrel glazing in curtain wall construction.

4. **D**   I is correct: While first cost is not the primary concern of life cycle costing, it is one of the economic factors considered. II is correct: The cost of maintenance is part of the evaluation. III is correct: The durability of a product or system is considered in the cost of repair and is part of the overall evaluation. IV is correct: The comparison of product or system life is one of the factors evaluated in life cycle costing.

5. **B**   Conventionally laid wood strip floors are blind nailed in order to conceal the nails and tightly drive each strip closer to the preceding strip. At the wall, however, there is no room to drive a nail at a 45-degree angle. The final strip, therefore, is generally face nailed. It is also set about ½ inch away from the wall in order to allow for expansion. This space is usually covered by the wood base.

6. **D**   The standard slump test measures the workability of concrete. The slump of a standard cone of fresh concrete is measured, and the stiffer the mix the less the slump.

7. **B**   The finish of ferrous metals is of particular concern, as steel is susceptible to corrosion when exposed to air and moisture. Under these circumstances, alkyd-based paint is preferred over oil-based because it dries faster, harder, and with greater resistance to the weather. Furthermore, it is generally advisable for all coats of paint to be of the same base type. Incidentally, coal-tar enamel (A) should be specified where steel is installed underground or in water. In that case, the primer should also be a coal-tar product.

8. **C**   The only true statement is correct answer C. The gravel fill in which the drain tiles are set (A) provides a porous material through which water may flow to the open joints between the tiles and then away from the building. The waterproof joint between slab and wall (B) provides a positive seal, as well as thermal insulation for the slab. It is sometimes omitted in temperate areas, but only when the temperature remains above freezing during the night, as well as the day. Finally, the capillary rise of ground moisture (D) should be effectively stopped by the waterproof membrane barrier; that is, in fact, its very purpose. Furthermore, capillary action rarely occurs in coarse granular fills, such as the one indicated in the detail shown.

9. **B**   In the control of surface groundwater, the finish grade should be set at least six inches below the slab level, as shown (D). However, the grade should be sloped away from the foundation at all sides (correct answer B), in order to protect the structure.

If the drain were set below the footing (A), there would be danger of undermining the footing from the flow of running water. Therefore, foundation drains are always placed as shown. Finally, the waterproof membrane should never be set below the granular fill (C), because the purpose of the fill is drainage, insulation from the soil below, and most importantly, the reduction of capillary action.

10. D   Iron did not become an important construction material until it could be produced in great quantity. When the Englishman Abraham Darby developed the blast furnace around 1750, mass production became possible. Prior to that time, iron was difficult to produce in quantity (C), and it was also unpopular because of its poor resistance to weather (A) and its lack of historical precedent (B). Iron ore was plentiful, however, and relatively simple to mine. At first, mass-produced iron was used for machines, utensils, and rails, and shortly thereafter, the first cast iron bridges were constructed in England.

11. A   Portland cement is manufactured in several types, each with characteristics that make it preferred for a particular set of conditions. For example, the low-heat cement of this question would be suitable where slow setting is desired and where the heat generated by ordinary cements would be excessive and lead to serious cracking. This situation often exists in massive concrete work (correct answer A). For general use (B) one would use Type I (normal). During very cold weather (C) one would use Type III (high early strength) in order to generate greater heat and offset low temperatures. In alkaline areas (D) the best choice would be Type V (sulfate-resisting), which is better able to withstand chemical attacks from sulfates found in the ground or in some building materials. While the

different types of cement have different rates of strength gain, they all eventually reach about the same strength.

12. C   Bulb tees are used in gypsum concrete construction as structural sub-purlins. The bulb tees are fastened to the primary framing and provide support for the form boards that receive the gypsum concrete. They also anchor the deck against uplift forces, restrict deck movement due to temperature changes, and provide lateral bracing for the roof structure. Gypsum concrete roof decks are lightweight, they set rapidly (30 minutes), and when used over gypsum or mineral fiber form boards, they are classified as noncombustible.

13. B   An unprotected, partially finished masonry wall that may be saturated by rain can cause problems. The wall may take months to dry out, and during this drying period, efflorescence may appear on the wall surface. If there is a threat of rain, the wall should be positively protected, as described in correct answer B. Incidentally, when the outside temperature drops below 40 degrees (C), no further masonry work should be performed unless adequate precautions are taken against freezing.

14. A   Type II N is considered noncombustible construction. The rest are considered combustible.

15. C   Choice I is correct: Designing with native landscaping is preferred to using exotic or imported plant types. Indigenous plants tend to survive longer, use less water, and cost less. Choice II is not correct: Placing any structure in a floodplain, even those that resist floodwater, is not desirable. Placing buildings in a floodplain can increase flooding further downstream. Choice III is also correct: Buildings sensitive to the benefits of solar orientation and passive and active solar gain techniques save energy and are more visually aligned with local climatic conditions. Choice IV is correct as well: In-fill development and proximity to a variety of transportation options are design principles that benefit the inhabitants and their environment.

16. **B**   Three exits are required where there are more than 500 occupants and four exits are required where there are more than 1,000 occupants. Other areas typically require only two exits, but where the path to the first exit exceeds the allowable travel distance, a third exit would be required.

17. **B**   Fire door requirements are covered in Chapter 43 of the UBC, as well as corresponding sections of other building codes. Class A doors (three-hour rating) are used in openings in fire walls or division walls between buildings or sections of buildings. Class B doors (1½ -hour rating) are used in openings in enclosed vertical shafts, such as stairs or elevators. Class D doors are also rated 1½ hours, but are intended for severe fire exposure in exterior walls, and no glass is permitted. Statements A, C, and D are true, while statement B, the correct answer, is false.

18. **C**   There are few definitions as clear as that for incombustible. It simply means incapable of igniting or supporting combustion when exposed to fire, usually at an air temperature of 1,200 degrees. An incombustible building material, therefore, is simply one that will not burn.

19. **A**   In the application of built-up composition asphalt roofing materials, the use of angled cant strips is standard procedure. The reason for this is stated in correct answer A. When 15# felts are bent at a 90-degree angle, they have a tendency to crack or rupture. The resulting weakness, when combined with the natural shrinkage and expansion due to weather and other movements of the building, may produce serious roofing failures. Cant strips are generally cut from wood or manufactured from a variety of lightweight, inexpensive materials.

20. **A**   I is correct: Solar shading, whether from landscaping or architectural elements, can regulate the insulation to increase winter light and reduce warm summer sunlight. II is also correct: Urban heat island effect is the tendency of a building roof to absorb solar radiation during the day and then emit heat radiation during the evening. Roof systems with grass or light-colored roofing material reduce the urban heat island effect. III is not correct: Sustainable design encourages approaches that reduce the area allocated to parking. IV is correct: The type, location, and size of building fenestration are a key aspect of architectural design for sustainable projects.

21. **C**   In the same order as the statements, laminated members are generally manufactured from short and narrow lengths of lumber. One of the great advantages is that these small pieces may be end-joined to form almost any length desired with no loss in strength. Concerning the thickness of lumber used, laminations rarely exceed two inches in thickness, but they may be thinner. In the case of some curved members, for example, where the bending radius is short, ¾-inch-thick laminations are often used. The next statement (C) is the correct answer. Simply-supported single spans of 60 feet or more are not uncommon. In the form of arches or simple trusses, clear spans may reach 100 feet or more in length. The last statement is false. The very best appearance grade is called "premium," which is specifically intended for use in exposed construction.

In descending order of appearance grade, next comes "architectural" grade and then "industrial" grade.

22. **D**   Objects that project from walls constitute a hazard to those with sight disabilities. Therefore, projections are limited to four inches when such hazards are mounted between 27 and 80 inches from the floor. When located below 27 inches, however, wall-mounted projections are easily contacted by a hand-held cane before the blind person reaches the object. Therefore, the ANSI standards permit these objects to protrude any amount, so long as the object does not reduce the clear minimum width of an accessible route.

23. **C**   The ancient Egyptians employed stone because it was available and its use resulted in stable and permanent structures. The immense stone blocks were difficult to handle (A), but the vast armies of unpaid laborers, prisoners, and slaves solved that problem, as well as the problem of cost (B). Finally, stone does have a high value of insulation appropriate for a warm climate (D), but that quality was incidental to the ancient Egyptians.

24. **D**   All of the statements are true with the exception of the last one. As the water-cement ratio increases, the strength of the concrete decreases. The densest and strongest concrete, therefore, is obtained by using a workable mix with the lowest water-cement ratio that will enable the mix to be thoroughly compacted.

25. **D**   All of the above. Building codes are intended to protect the public health, safety, and welfare.

26. **B**   Aluminum is highly resistant to corrosion from normal weathering because of the tough film of aluminum oxide that rapidly forms on exposed surfaces. It is subject to attack by alkalies, however, and must therefore be protected from contact with wet concrete, mortar, and plaster. Various temporary plastic coatings are used for this purpose. Aluminum is also subject to galvanic action (A), and therefore where it comes in contact with other metals, the meeting surfaces should be insulated by gasketing or by painting with asphaltic or bituminous paint.

27. **C**   The most important consideration in the selection of roofing nails is that they be non-corrosive, such as copper or aluminum (correct answer C). The danger in using steel nails (A) is that they may eventually rust, discolor the tiles, or even corrode to the point of failure. Annular or helical nails (B), whether manufactured from a non-corrosive metal or not, have great holding power, but strength of this magnitude is generally unnecessary. Elastic cement (D) is often used to seal a roof tile joint, but it does not have sufficient holding power for roof tile application.

28. **B**   A needle beam is used in the situation described in correct answer B. A hole is cut into an existing masonry wall, through which the needle beam is placed at right angles to the wall. Each end of the beam is supported at some distance from the wall and serves as temporary support while the required opening is cut. After the new lintel or arch is formed, the needle beam is removed and the original hole is patched. Needling may require more than one temporary beam if the opening is very wide.

29. **A**   Non-load-bearing partitions are made from either solid or cored gypsum blocks that are 12 inches high, 30 inches wide, and from 1½ to 6 inches thick. They are generally set on a waterproof base course and joined by a mortar consisting of one

part neat, unfibered gypsum to three parts sand (correct answer A). Portland cement mortar (B) is used for brick or concrete block construction and is classified as type A, B, or C (answer D), depending on its resistance to weather exposure.

30. **A**   A number of patterns and dimensions of trim are produced in a variety of hard and soft woods, for both interior and exterior use. Each was originally developed for a specific appearance when covering the joint where two surfaces meet. At a wall and ceiling intersection, for example, a crown mold is often used (correct answer A). Casing molds are used around openings, corner beads are used at exterior corners, and chair rails are applied at the midheight of walls to protect them from furniture damage.

31. **D**   The door label designating a 1½-hour fire-resistive rating is used in class B openings, generally in vertical stairwells. For that designation, the fire assembly (door frame, hardware, etc.) is required to be self-closing. This means that it is equipped with a mechanical closer that insures closing and latching after being opened. An automatic closing device, on the other hand, is used on a fire assembly that may remain open, but will close automatically if subjected to an increase in temperature. This is required for a three-hour fire-resistive rating. Glazed openings in a class B assembly are permitted, providing they do not exceed 100 square inches. The correct answer therefore is D, as neither statement applies to the door label described.

32. **B**   The amount of water vapor that the air can hold increases with the temperature. Within a building, this moisture can come from mechanical equipment, perspiration, and even breathing. The amount of moisture held in the air is usually expressed as relative humidity. The temperature at which the air becomes saturated and vapor condenses is known as the dew point. Although air in a room is generally well above its dew point, the temperature of some surfaces is not, and that is why we usually see condensation first appearing on the inside of windows, which may be cold. Because vapor flows from high to low pressure, it can pass through porous construction, cool to its dew point, and condense as water within the structure. This can be prevented by the use of a vapor barrier. The barrier itself, usually a plastic film or metal foil, must be at a temperature above the dew point, and therefore it is always installed on the warm or room side of the construction beneath the interior finish (correct answer B).

33. **B**   I is correct: LEED has several options for improving Indoor Air Quality (IAQ) including filtering the air system and installing low Volatile Organic Compound (VOC) paints and caulking. II is also correct: Methods to store, recirculate, and locally distribute rainwater are encouraged. III is correct: Innovative solutions to energy conservation, such as fuel cells, photovoltaic panels, and gas turbine energy production, are encouraged in the LEED accreditation system. IV is incorrect: Unfortunately, the LEED system awards no points for designs with strong aesthetics.

34. **C**   The standard sizes of metal casement windows are all based on brick masonry modules (correct answer C), as these windows were originally developed for use in brick wall construction.

35. **D**   Intumescence is the formation of a charred, foamlike material on a coated surface that has been subjected to intense heat. Paints that are able to sacrifice

themselves and form this char are called intumescent paints (correct answer D). These special paints are formulated to protect underlying flammable surfaces by insulating them from the intensity of heat, thus delaying ignition. Generally, the fire is not extinguished, but merely retarded. However, delaying the flame spread by even 15 to 30 minutes may save many lives in the evacuation of a school or hospital. Concerning the other choices, other coatings are available that satisfy these special requirements.

36. **B** Perret's eight-story apartment building, built in 1903, was the first to use reinforced concrete frame construction. Although the concrete columns and beams were sheathed in terra cotta, the structure was revealed to a degree unprecedented in domestic architecture. The infilling panels were ceramic tile decorated in relief, and the building's glass areas were more extensive than usual for that time.

37. **A** What is shown is an expansion-type dividing strip, the purpose of which is described in correct answer A. This type of strip is manufactured from a variety of expandable materials, which are used wherever such joints are required. The joint width depends on the total anticipated structural movement.

38. **D** Blisters may be caused by all of the factors listed (correct answer D). Most of the moisture comes from exposure during construction or from poured roof decks that have not had enough time to dry adequately before the application of roofing. Blisters may also result from failure to cement the felts securely to each other, to the insulation, and to the deck, all of which create air bubbles.

39. **C** A concrete masonry unit is specified by its width, height, and length, in that order. The actual dimensions are generally ⅜ inch less, to allow for a mortar joint. The correct dimensions are therefore shown in answer C.

40. **A** Of the types of paint in this question, only acrylic latex (correct answer A) is completely solvent-free, and therefore presents no flammability hazard in either storage or application. Polyester-epoxy combines the physical toughness, adhesion, and chemical resistance of epoxy with the color retention and clarity of polyester. Alkyd paints are made by combining synthetic materials with various vegetable oils to produce clear, hard resins. Polyurethane contains flammable solvents, but is highly resistant to abrasion, impact, and chemicals.

41. **B** The Crystal Place was designed by Joseph Paxton to house the Great Exhibition of 1851 in London. Although it employed cast iron framing (I), and exhibited great spans (II), it was not the first building to do so, as it was preceded by several bridges, factories, and warehouses. It was, however, standardized, prefabricated, and erected in only three months (III); and in 1851, it was the largest building the world had ever seen (IV), enclosing nearly a million square feet. Finally, the Crystal Place was not fireproof (V), and in fact, the entire iron and glass structure was totally destroyed in a rapid fire in 1937.

42. **D** All of the statements are true and therefore D is the correct answer.

43. **C** The decking shown and the concrete floor slab placed over it act together as a composite member (correct answer C). This type of action is made possible by the

deformations in the deck, which function as mechanical connectors between the slab and deck.

44. **A**   Handrails and grab bars that are required by the ANSI handicapped standards must have a diameter of gripping surface between 1¼ and 1½ inches. When they are mounted adjacent to a wall, the clear space between the handrail or bar and wall must be no less than 1½ inches (correct answer A). In addition, the gripping surfaces must be continuous and free of any sharp or abrasive element.

45. **A**   Although all of the choices are valid and might be implemented to reduce the heat loss in this old building, the most important first step is to reduce the U factor in that portion of the building envelope that constitutes 85 percent of the exterior wall surface. In this case, applying insulation and gypsum board to the interior surface of the exterior walls will reduce its U value from approximately .56 to .24 (correct answer A). Similarly, double glazing (B) will reduce the U factor from 1.13 to .45; however, it affects only 15 percent of the building envelope, and therefore it would be a secondary consideration. Caulking at windows and doors (C) as well as re-grouting the masonry walls (D) will reduce or eliminate leakage, or infiltration, and these are additional considerations in reducing heat loss.

46. **C**   No excavation for the new footings should be started until the footings of the existing building are first extended down to the level of the proposed building's footings. This is accomplished by using temporary supports, such as timber shores or steel needle beams, to carry the weight of the existing building until the new and lower foundation is in place. When the entire weight of the existing wall is transferred to the new section of the foundation, the shoring and underpinning are removed and the foundation for the new building may proceed. The new work is generally performed on all sides simultaneously.

47. **B**   The L-shaped extrusion forms a continuous condensate gutter (correct answer B). Under certain temperature conditions, moisture vapor, or humidity, changes to liquid through condensation. Generally, moisture vapor flows from high to low pressure independent of air flow. The formation of condensation is particularly prevalent during winter months on glass surfaces that are heated on one side. In a skylight, this moisture must be removed by condensation gutters or else the moisture may cause damage in other parts of the structure.